The Planets

Some Myths & Realities

A HALSTED PRESS BOOK

The
Planets

Some Myths & Realities

Richard Baum
With a Foreword by Patrick Moore

JOHN WILEY & SONS
New York

Library of Congress Cataloging in Publication Data

Baum, Richard Myer.
　　The planets: some myths & realities.

　　'A Halsted Press book.'
　　Bibliography: p.
　　1. Planets. I. Title.

QB601.9.B38　　　　　523.4　　　　73-7583
ISBN 0-470-05930-3

Published in the USA
by Halsted Press, a Division
of John Wiley & Sons Inc.,
New York
Printed in Great Britain

For
Audrey, Julian
Adrian & Jacqueline

Contents

Contents

List of Illustrations

Foreword

ASTRONOMY today is entirely different from the astronomy of even ten years ago. Men have reached the Moon; probes have been sent out to Venus and Mars, and sophisticated new equipment has been developed to help in studies of the remote stars and star-systems. Each year brings its quota of new developments, and progress in all fields of astronomical science is more rapid than ever before.

Inevitably, much has been written about all these advances, and equally inevitably there has been a considerable amount of duplication. It is not often nowadays that one finds a book which is intelligible to the general reader, and yet has a theme all its own. Richard Baum has provided such a book, which is why I am particularly glad to have been asked to provide a Foreword for it.

The history of astronomy is a fascinating study, if only because there are so many aspects of it and so many intriguing characters are involved. Sometimes there are curious contradictions; how can we reconcile Sir William Herschel, the 'explorer of the heavens' and discoverer of the planet Uranus, with Sir William Herschel, who firmly believed that intelligent beings lived in a cool region below the hot surface of the Sun? But quite apart from this, there are remarkable episodes which in some cases have never been satisfactorily explained, and which have seldom or never been described in full.

This is the theme of the present book. It does not set out to be a comprehensive survey of Solar System astronomy; there are plenty of works covering this ground. What Richard Baum has

done is to take various episodes of special interest and contro-
versial aspect, and describe them in a way which cannot fail to
intrigue all those who are astronomically-minded. Has the Moon
a satellite—a junior companion? What is the truth behind the
reported peaks on the surface of the planet Venus, towering up
to immense heights above the cloud-covered surface of that
extraordinary planet? And why did so many skilled observers, of
unquestioned integrity, report ring-like features associated with
the remote worlds Uranus and Neptune?

These are some of the topics with which the book is concerned.
Richard Baum, whom I have known well for many years, is well
qualified to tackle them. He is himself an observer of great
competence, so far that he has first-hand knowledge of the worlds
in the Solar System; he is also a skilled astronomical historian,
so that his researches have been thorough and accurate; and he
has the ability to write in a vein which will do justice to the
subjects with which he is concerned. A great deal of labour and
correspondence has gone into the preparation of the book, and
he has produced a work which will appeal both to interested
onlookers and to serious students—both professional and
amateur—who are undertaking their own research into historical
astronomy. To the best of my knowledge, the episodes described
here have never before been collected in book form, and some of
the material has not previously been published anywhere. I am
confident that all those who read it will be as fascinated as I was
when I first read the book in manuscript form.

PATRICK MOORE
Selsey 1972

Introduction

TELESCOPIC study of the planets is naturally subject to and dependent on several factors: the physiology and complex psychology of the observer, the quality of the optical system employed, the vagrant moods of our atmosphere, the agitations of which prejudice any result, and the apparent size and optical properties of the object under observation. In each of these repose malign attributes, as it were, which not only conspire to thwart the observer in his purpose, but sometimes induce subtle optical effects that may deceive even the most experienced observer. Vigilance, experience and skill militate against them, but in the early days of planetary astronomy, as with the earliest voyages of terrestrial discovery, ignorance combined with novelty produced reports of things half-seen that shimmered uncertainly in the dubious borderland between the known and the unknown.

Inscribed in the annals of planetary observation we locate the record of unusual observations, not confirmed by later observers, and of which we can offer no satisfactory explanation; they are minor side-issues—footnotes to the familiar. In part they represent a vast corpus of abandoned workings along the periphery of the known; divorced from the mainstream of astronomical progress, and by virtue of their character, largely forgotten. It is with this type of observation I am here concerned. And in the following pages I have set out to describe a few such reports. Sometimes they are odd, sometimes diverting. Sometimes they pose a distinct possibility; but always they are interesting, and often instructive.

In some respects we may draw a parallel with the legends derivative of historical geography. The persistent mention of a huge mountain at the south pole of Venus, for instance, reminds us of the schematic geography of the Arctic basin compiled by the Franciscan minor friar, Nicholas of Lynn. Having sailed north from Norway about 1360 A.D. he apparently reached the shores of southern Greenland. On his return he prepared a chart in which he portrayed the Pole as a shining black magnetic rock standing in the midst of a giant whirlpool, the whole encircled by prodigious ramparts and ice-covered mountains, divided by numerous channels through which the sea rushed with incredible violence towards the whirlpool. A fanciful picture perhaps, but less so than some that were advanced during the eighteenth and nineteenth centuries (Edgar Allen Poe utilised this concept, incidentally, in his story *MSS. found in a Bottle* (1840)).

Again, false planets have their counterpart in lost or mythical continents, and so on. Percival Lowell frequently prefaced his popular writings on Mars by equating the astronomer with the explorer. The analogy is apt. No matter that one is relatively static, in the physical sense, and the other mobile; each feels a strong compulsion to interrogate the unknown in search of new facts about his environment.

Rather than undertake an analytical treatment, I have adopted a casual, though not indifferent, approach. With each subject I have taken the facts as I see them and have attempted to weave them into one continuous story. It is required in the history of any subject that the past shall speak for itself, and this belief has had a strong influence in the arrangement and preparation of the book. Inevitably I have quoted heavily from original sources, inserting comment and opinion, which are my own, where these appear necessary to maintain continuity. In all but one instance, the subjects lend themselves to simple historical exposition in the order of unfolding events, but in the discussion on hypothetical satellites of the Moon I have found it necessary to preface and clarify the historical narrative by some explanation of the peculiar difficulties that impede the progress of a search for such bodies.

Except in the case of 1921e, the setting of each of these histories is the nineteenth century; one of the most brilliant periods in the history of the science, when older departments were boosted by the injection of new concepts and instrumentation, sub-divisions were widened and systematised and new branches created. In the first and second chapters, which deal respectively with the quest for satellites of the Moon, and the fabled mountains of Venus, I have been obliged to bring matters forward from the prescribed century, and with the latter to refer back to an earlier period, chiefly because here the interest posed has a relevance to modern research. With regard to the aforesaid mountains, the events of the past decade or so, namely the innovations of radar astronomy, have been more summarily described because of the difficulty of achieving historical perspective at such close quarters; reference being made only to the main happenings, and to some of the principal trends and influences.

Eight case histories are discussed. The first deals with hypothetical satellites of the Moon, an idea briefly revived by the demands of the Space Age. In the second, I have tried to give a general description of the origins and foundation of the Venus mountain hypothesis, once thought to be no more than a romantic survival but now strangely recalled by the discoveries of radar astronomy. With three and four, the mystery is total. Five and six decidedly refer to some optical aberration, though the circumstances are distinctly odd. Seven is concerned with a very strange error, with even stranger consequences. Again the lost planets of 1831 and 1835 cannot be explained except possibly by error.

The notes and references and the bibliography at the rear of the book signify the extent of my debt to those on whose published work I have so much depended. Of the works consulted many are not generally accessible, nor easily obtained, consequently I owe a debt to many for their help in providing answers to many questions, and for tracking down obscure references and pointing out others. Accordingly I wish to acknowledge the kindly help of Professor Clyde W. Tombaugh, Department of Astronomy,

New Mexico State University, and Dr Bradford A. Smith, who supplied information on their lunar satellite search, and willingly gave permission to reproduce two of their photographs taken at that time. Also I must thank Charles F. Capen, Jr, now of the Lowell Observatory, for his assistance in the same connection.

Whilst I have mainly relied on printed materials, unpublished letter sources have been used and quoted. For this I am indebted to Professor R. O. Redman, FRS, Director of the Cambridge Observatories, who made available letters from the Reverend W. R. Dawes and William Lassell to Professor James Challis on the question of the Neptunian ring, from the departmental archives.

For his invaluable assistance on this same topic, I am especially indebted to Dr David W. Dewhirst, Cambridge Observatories. I must also thank Dr E. W. Maddison, Librarian of the Royal Astronomical Society, for making available much data that would otherwise have escaped my notice. Again I must thank Charles F. Capen, Jr, for providing data of his personal observations of Neptune, likewise to Professor Ronaldo R. de Freitas Mourão, National Observatory, Brazil. For their attempt to clarify the elusive Bond observations of the Neptunian ring I am grateful to Professor Dennis Rawlins, Baltimore, and to Dr Joseph Ashbrook, editor of *Sky and Telescope*. I must also record my gratitude to Dr R. M. Goldstein and Dr H. Rumsey, the Goldstone Tracking Station, California, for sending a preprint of their paper *A Radar Snapshot of Venus*, and for giving permission to reproduce the radar maps of the planet which appear in the book; also to Dr James I. Vette, Director of the World Data Center, Goddard Space Flight Center, in making available photocopies of numerous papers on radar mapping of Venus, which provided much important background information for Chapter Two.

I must also thank William Beetles for having read the MSS, and for his many helpful suggestions; also William Wilson. And above all, Patrick Moore for much assistance over a long time. I must acknowledge the help given by T. Stanhope Sprigg of David and Charles for advice during the early stages of the book, and finally I must not forget to mention my mother and father for their help and encouragement, my wife for her gentle insistence and practical aid, and lastly Julian, Adrian and Jacqueline for their unbelievable patience, and for always honouring my solitude.

RICHARD BAUM

Chester 1972

1 The Search for a Satellite of the Moon

There is no dynamical reason why the Moon should not now have a satellite revolving about it at a distance of a few thousand miles from its surface.

Forest Ray Moulton (1931)

FOR brief periods in recent history, the Moon attained planetary status by its acquisition of a succession of minor companions. Whilst these were of an artificial character, the event raises the old question: Has the Moon a natural satellite?

When Professor Forest Ray Moulton admitted this possibility in 1931, he further observed that if the Moon once formed part of the Earth, as Sir George Howard Darwin inferred from his researches into tidal theory, then clearly it could not have held a satellite during its primordial phase and would not have one now.[1]

There is, however, another side to the coin as Professor Edward Charles Pickering (1846–1919), Director of the Harvard College Astronomical Observatory, proposed in 1888.[2] Numerous solid bodies, of diverse size and negligible mass, invest interplanetary space. Subject to the immutable laws of planetary motion, these fragments, the meteoroids as they are known, whirl unceasingly about the Sun in all directions and involve its retinue in a complex of intersecting orbits. Evidence of their existence may be had any clear night in the abrupt flare of a meteor.

Invoking the capture hypothesis, Pickering suggested that if one of these bodies happened to pass near the Moon with a moderate velocity, and in such a manner that the disturbing force upon it equalled the attraction of the Sun and counteracted it almost completely, then conceivably the meteoroid might be retained by the Moon as a satellite. Admittedly this is a possibility, though somewhat remote. According to the circumstances of the encounter such a transformation might take place at once, or in two or more stages, but the combined attraction of the Sun and the Earth would ever jeopardise the stability of the system thus created. Much depends on the original velocity of the meteoroid, and the direction from which it approached the Moon. If, for instance, its velocity were accelerated as it came within range of the Moon's attraction, it might be expelled from the system on a hyperbolic orbit and never return.

It may be remarked that any general formula capable of taking into account all the sequences involved in the capture of a small body and its subjugation to the status of a lunar satellite would have to be incredibly complex. The intricacy of all the possible motions to be considered, not the limitations of mathematical analysis, is the problem.

The suggestion is novel and attractive, if highly speculative. The statistical probability is small, but the nature of the proposition is such as to invite rather than deny conjecture, perhaps even profitably: a fact that excited attention late in the nineteenth century, and prompted E. E. Barnard to admit in 1895 '. . . it is a point that has sufficient plausibility about it to suggest a photographic search'.[3] Early in the 1950s, when official sources were concerned about the potential hazards of interplanetary space, the Office of Ordnance Research, US Army, reflected the same belief, and in an attempt to clarify the situation commissioned Professor Clyde W. Tombaugh to undertake a photographic search of the lunar satellite environment. It was thought, not unreasonably, that in its incidence a lunar satellite not only posed a distinct threat to the security of a vehicle travelling in the neighbourhood of the Moon, but also suggested a possible hindrance to tracking and orbital procedures. As Tombaugh found, these fears were entirely groundless.

This was no more than expected. Astronomers have had the Moon under almost constant surveillance since the dawn of telescopic astronomy, but at no time had anyone reported an object remotely compatible with its being a satellite. Some observers alleged moving specks near the Moon, but these might well have been anything from meteors to birds, dust particles or airborne seeds and the like so placed in relation to the observer as to appear momentarily connected with the Moon; there are many such line-of-sight observations on record.[4] But we digress. Whatever their cause, they could scarcely have been satellites of the Moon. What then is the prospect?

EXTENT OF SATELLITE ENVIRONMENT

In 1887 Willliam Henry Pickering (1858–1938), Harvard College Observatory, as part of E. C. Pickering's discussion, adopted two frames of reference in order to determine the maximum possible distance at which the Moon can retain a satellite. He based his first approach on a suggestion by H. L. d'Arrest that a satellite more than 70 minutes of arc distant from Mars would take longer to revolve about that body than Mars itself took to revolve around the Sun, and so would remain continuously in line between Mars and the Sun. Under these conditions the centrifugal force of the satellite, although exactly balancing the attraction of its primary, would be inadequate to resist that of the Sun, so that it would move out of range of the planet's influence, take up an independent orbit of shorter period than the latter, and circulate about the Sun as a faint planet. For his second approach Pickering quoted the example of the Earth-Moon system by noting that a satellite might still be held by its primary even though its distance from the planet puts it in a region where the attraction of the Sun predominates.

By a mathematical demonstration, Pickering showed that the true value for the maximum possible distance at which the Moon can hold a satellite lies between the results given by these two methods. Taking into account the effect of the Sun, his equation for the critical distance gave 37,000 miles, which in terms of angular measure gives a maximum elongation of 9° 47′ from the Moon.[5]

In accordance with this figure, Professor Tombaugh has calculated the theoretical motions for satellites with circular orbits out to a distance of 30,000 miles. Table 1 has been adapted from his results. Any observer who attempts a photographic search should consult the original table, published in 1959, as it also provides us with details of the Tangential Trailing and Dilution factors for the relevant distance/velocity relationship as referred to the 1956 search at Flagstaff.[6]

TABLE 1
Theoretical Motion for Lunar Satellites with Circular Orbits
(after Tombaugh)

Luni-centric distance (miles)	Period (minutes)	Miles (a minute)	Tangential angular velocity (a minute)
1,500	178	52·9	47″·1
2,000	274	45·9	40″·9
3,000	504	37·5	33″·4
4,000	775	32·4	28″·8
5,000	1,080	29·0	25″·8
7,000	1,790	24·6	21″·9
10,000	3,060	20·5	18″·3
15,000	5,620	16·8	15″·0
20,000	8,650	14·5	12″·9
25,000	12,100	13·0	11″·6
30,000	15,900	11·9	10″·6

CONSIDERATIONS ON VISIBILITY

Suppose the Moon furnished itself with a companion. What chance has the amateur with his modest equipment of effecting its discovery? This naturally depends on several things, but in the main on the size of the body concerned. As a satellite in excess of half a mile in diameter would quickly attract attention, the question has regard to a discrete particle well below the naked eye threshold. The Harvard search of 1888 had a limiting magnitude of 10. To find out just how large an object this implied, W. H. Pickering adopted three lines of inquiry:

(1) It had been demonstrated that the most probable value of the stellar magnitude of the Sun was −25·5. Assuming the brightness of the Sun to exceed that of the Moon when full by about 600,000 times, Pickering expressed the stellar magnitute of the full Moon as −11·1. On this basis he thought that if the Moon had a small companion of magnitude 10, it would give out 1/275,000,000 the light of the full Moon, and so have a diameter of 1/16,600 that of its primary, or about 687 feet.

(2) Some years earlier when Mercury had been near its greatest elongation from the Sun, it had been compared with Procyon, which lay near. This revealed them to be of practically the same brightness, 0·5. Taking Mercury at its mean distance from the Sun, Pickering computed that if placed at the distance of the Moon it would receive only 0·15 of the light. Using Lambert's formula he found that a body of the same angular diameter as Mercury, if situated at the distance of the Moon from the Sun, and at full phase, would yield 0·47 the light of Mercury under those conditions, or be equal to a 1·3 magnitude star. At the distance of the Moon from Earth, such a body would be about 8 miles in diameter, from which Pickering thought that a satellite of magnitude 10, of similar albedo to Mercury, would have a diameter of approximately 754 feet.

(3) For his final approach Pickering adduced a photographic result which gave a smaller value. He took several photographs of Polaris, throwing its image out of focus so as to cover an area of from 1·5 to 3·7 millimetres in diameter. The brightness of these images was then measured and compared with that of the full Moon, and found to be on average 1/665,000 as great. This gave the stellar magnitude of the full Moon as −12·35, and the diameter of a satellite of magnitude 10 as roughly 337 feet.

From the average of these results Pickering deduced that a lunar satellite of the tenth magnitude would have a linear diameter of about 592 feet, or 0″·1 at the mean distance of the Moon.[7]

Visual detection of such an object, even if difficult, would not prove impossible, and falls well within amateur capability. Since the Moon is perhaps the most frequently observed celestial body, we concede that in all probability Pickering's hypothetical satellite would not pass unnoticed. It might be seen in transit on the lunar disk, especially just beyond the terminator in the dark of the Moon, or just be detected by someone watching stellar occultations.

As a satellite only shines by reflected sunlight, it follows that its light will increase and decrease in exactly the same proportion as that of its primary, the Moon. This means that it will be brightest at full Moon, and thus quite conspicuous. But this is not so. At this time the sky is flooded with light some 10,000 times more than on a moonless night, hence the assumed satellite would be consumed in the glare of its primary, or at least rendered so faint as to be exceedingly difficult to pick up. Tombaugh states that the detectability factor for a dim object under these conditions is reduced by about ten stellar magnitudes.[8] This is caused by atmospheric scattering, which is obviously more intense in the direction of the Moon, a fact easily demonstrated any clear evening when the Moon is at or near full phase.

Quarter phase offers a slightly better opportunity. Then moonlight is not too troublesome, whilst the solar illumination on a satellite is enough to raise its light to within two magnitudes of greatest brilliancy.[9] Thus at this phase Pickering's hypothetical tenth magnitude satellite would descend to magnitude 12, and be correspondingly less easy to find.

Accidental discovery of such a body, though possible, is none the less improbable. The nature of lunar observation militates against the likelihood. Visual study of the Moon usually consists of improving knowledge of surface terrain and conditions, and is normally conducted at high magnification, but this implies light loss and a narrow sphere of operation. Effectively, this means that at any time interest is concentrated on a relatively minute portion of a circle only $\frac{1}{2}°$ in diameter, since this is the apparent angular diameter subtended by the Moon to the unaided eye. As the satellite environment describes a circle 20° in diameter around the Moon, according to W. H. Pickering, it is obvious that the region where a satellite is most likely to be found is never explored or even scanned. When it is, as in the case of stellar occultation work, attention rarely extends more than a degree away from the limb of the Moon.

Exceptionally, there have been occasions when general surveys formed an essential part of the study in hand. Late in the 1930s, and during the early 1940s, teams of American observers under Walter H. Haas and Dr Lincoln La Paz systematically swept the lunar disk and its environs for evidence of meteoritic phenomena (impact flares on the surface and possible lunar meteors).[10] More recently, other groups have elected to regularly patrol the surface for traces of residual activity; an investigation organised on an international basis in connection with the early Apollo missions under the code name LION (Lunar International Observers Network).[11] But as Tombaugh again remarked in his 1959 report, such observations hardly qualify, because they are invariably made with telescopes incapable of detecting very faint objects in bright moonlight.[12]

Taking into account all the peculiar difficulties attached to this problem, but acknowledging that some observers have better eyesight than others, it is supposed that a satellite of the tenth magnitude marks the limit of discoverable object accessible to the type of telescope commonly employed to observe the Moon. Most probably a satellite would be exceedingly tiny and beyond the capabilities of modest instruments, and we would play a long game of celestial hide and seek before it was finally caught.

It is understood then that if the Moon has a small companion, the existence of such would only be disclosed after a deliberate and conscious effort to search it out with large and sophisticated equipment.

TECHNIQUES AND OPPORTUNITIES

Two methods are available to an earthbound observer who proposes to search for a lunar satellite. Either he may sweep visually, or else carry out an exhaustive photographic survey of satellite space.

The faintness of the assumed object, the adverse influence of moonlight, the inconveniently large area of sky to be examined, and the large number of objects to be checked for motion (the only feasible criterion of identity), impose severe limitations on the comparative effectiveness of the visual method. Tombaugh believes it has a reasonable chance at quarter phase when Moon glare is less intense. According to his estimate a 12-inch telescope, which has a theoretical limiting magnitude of 14, appears quite adequate to probe the inner satellite environment out to a distance of 3,000 miles from the Moon. Here the tangential or proper motion ranges from 56 to 33·4 seconds of arc per minute. For distances in excess of 3,000 miles where the tangential motion is proportionately less, ranging from 33·4 to 10·6 seconds of arc per minute, Tombaugh considers a 24-inch telescope absolutely necessary to supersede the photographic limit to date.[13]

Whatever its orbital inclination, a satellite will almost certainly be seen at some time or other, if not regularly, as a moving speck of light projected against the rugged backcloth of its primary. If close to the Moon it will move rapidly; if distant, more slowly; but a minute or so of watching will expose an appreciable shift in position. A satellite with retrograde motion should also be borne in mind. Transits are worth looking for at quarter phase on the unillumined side. Low powers and displacement of the bright hemisphere from the field of view will, of course, be necessary.

It is readily observed that everything in this matter points to the use of wide-angle photography. This removes many of the restrictions that inhibit the visual technique. But the light of the Moon is still a major obstacle. Long exposures are impracticable at large-phase angles when the satellite is comparatively more conspicuous. For smaller angles the difficulty is diminished. Then, however, the satellite would be correspondingly reduced in light, and the problem increased.

To be rid of Moon glare is thus the chief requisite to the institution of a successful photographic search for a satellite of the

Moon. Such an opportunity is afforded during a total eclipse of the Moon. The almost complete exclusion of light from the face of the Moon at this time provides an excellent opportunity to discover a faint body within its immediate neighbourhood. Even so, the amount of sunlight that infiltrates into the shadow by refraction is usually considerable. Ideal search conditions prevail when the cloud content of the terrestrial atmosphere is high around the base of the shadow; the Moon is then practically invisible.

Sky illumination is too great whilst any part of the Moon remains in the penumbra, but after its total entry into the umbra the sky light is subdued to that of a moonless night. As the umbral shadow is roughly $1°·5$ in diameter, and since the penumbra but $0°·5$ beyond, a search for a satellite in full phase can be made to within about 2,500 miles of the lunar surface. This may be extended to within 1,000 miles of the Moon where the partial illumination on a satellite amounts to approximately 50 per cent at the start and end of totality.[14] Thus,

> If the region within which a satellite of the Moon might occur is photographed during a lunar eclipse upon a series of plates, the presence of a satellite outside the limits of the Earth's shadow could be detected by comparison with star-charts of the same region. An easier test would be afforded by the relative motion of the satellite and the stars. Whatever the adjustment of the telescope, the satellite would in general describe a trail having a different length and direction from that of the stars, and could thus be immediately distinguished from them. This method would fail if the relative motion of the satellite was exactly equal and opposite to that of the Moon, but the probability that this would occur would be very small.[15]

There is some chance of a satellite actually reposing in the Earth's shadow cone during totality, but as Tombaugh further remarks, 'The probability of a lunar satellite lying within the Earth's shadow during a given lunar eclipse decreases with increasing distance from the Moon.'[16]

Unfortunately, dilution losses caused by image trailing are unavoidable, as the relatively short duration of totality does not allow sufficient time to take photographs at variable rates of driving. Such losses will naturally be heaviest when the satellite is travelling at right angles to the line of sight, and of course will become more evident as the apparent distance between the satellite and the Moon lessens. Driving the camera at a constant speed, and guiding on some prominent surface feature on the Moon, appear to be the best strategy. In this way, the narrow zones either side of the Moon where the satellite will be at its greatest elongation are well favoured. These are 3° wide, a total of 6° out of the 360° circumference of the satellite orbit. In this position, the satellite would be moving directly towards or away from the observer, and since its motion would then coincide with that of the Moon, it would register on the photographs as a point image, and so quickly distinguish itself from a multitude of star trails.

HARVARD 1888

Edward Pickering decided to test his discussion in 1888. 'The recent improvements in astronomical photography,' he subsequently wrote, 'permit this problem to be attacked in a more satisfactory manner; and we may expect to discover such a satellite, if it exists, or to prove that no such object above a certain limit of brightness revolves around the Moon.'[17]

The experiment, ostensibly the first of its kind, was scheduled to take place during the total eclipse of the Moon on January 28 1888:

> ... with the aid of the Bache telescope. The object-glass of this telescope is a photographic lens by Voigtländer, corrected by Alvan Clark and Sons, and described and figured in the Memoirs of the American Academy, Vol. XI., p. 184. Its aperture is 20 centimetres, and its focal length 115 centimetres. A region 5° square is very satisfactorily covered by this lens, and the images of the stars are moderately good over a region 10° square.[18]

This instrument was detached from its normal pier, 'and mounted on a platform of plank frozen to the ground, where a nearly unobstructed view to the northeast was obtainable.'[19]

For two years, the Bache telescope had been used by Willard P. Gerrish to photograph selected star fields for the Henry Draper Memorial work on stellar spectra. Gerrish had joined the observatory staff in 1886. His remarkable aptitude for improvisation quickly caught the eye of Pickering, and he had been assigned the photographic duty of the Draper Memorial studies. It was natural, therefore, that the eclipse photography should be given into his charge. F. E. Fowle, another assistant, was detailed to develop the plates as soon as possible, and pass them to Professor Pickering for immediate checking, so that if a satellite was suspected, its position could be measured with the 15-inch refractor, and the plate repeated.

Gerrish arrived at the observatory before dusk on the night of the eclipse. The air was chill and the sky littered with broken cloud. Conditions were generally disturbed and the seeing inferior. To the north-east the Moon hung expectant in Cancer, somewhat east of the loose cluster Praesepe. Pickering reckoned that its apparent hourly motion in right ascension during the eclipse would amount to 140 seconds of arc, and 160" in declination:

Accordingly, if the satellite had the same apparent motion as the Moon, it would describe a line upon the plate at the rate of about 30" per minute, if the telescope was adjusted so as to follow the motion of the stars. If the satellite was faint, the diameter of its photographic image would not exceed 10". It could traverse this distance in about twenty seconds of time, and a longer exposure would make its image a line of greater length without enabling a fainter object to be photographed than could be taken in the first twenty seconds. It therefore seemed better to make the telescope follow the Moon in right ascension, by increasing the speed of the driving clock so that it would gain 140 seconds per hour. In this case the stars would form short lines parallel to the equator, and the satellite a still shorter line perpendicular to them. This motion of the satellite would thus be about one-twelfth of that of the star, and it would therefore produce an image equal to that of a star of twelve times its brightness. This effect may be still further increased by pointing the axis of the telescope to the west of north, as was actually done in making the photographs here described. Allowance was thus made for a portion of the motion in declination. The actual difference was less than that here computed, perhaps owing to the differential refraction or other causes. The apparent motion of the Moon was about one-third of that of a star, and in a direction at right angles to it. The motion of the satellite has here been assumed to be the same as that of the Moon.... A satellite 1° distant from the Moon might have a relative motion with regard to it of 30″ a minute.[20]

With these objects in view, total coverage of lunar satellite space 'especially in the direction of the ecliptic, where it seemed more probable that a satellite might be expected,' was anticipated.[21]

Shortly after dusk the search commenced. As twilight was too strong for an exposure of the usual length, Gerrish made two exposures of a ¼ second and two seconds with an interval of a minute and a half between; this gave him chance to test the running of the telescope drive.

Gerrish obtained his best photograph at 11h 15m, GMT, after a nine-minute exposure. Fixed on RA 8h 40m, Dec, +18° it gave satisfactory coverage of satellite space.

The search began at 10h 29m, and ended at 12h 44m GMT. Twenty-four plates, each measuring 8 × 10 inches, and all but three the most sensitive obtainable, were exposed, and embraced a region 30° square centred on the Moon in RA 8h 40m, Dec +18°. Exposure times ranged from a quarter second to ten minutes. Photographs with RA 8h, 8h 40m, and 9h 20m, for each band of declination +8°, +18° and +28° were planned to blanket the entire satellite zone, and those taken between 10h 41m, and 11h 26m, had precisely this mission. Unfortunately Gerrish committed an error in hour angle setting at 10h 51m, and centred a plate in 9h instead of 8h 40m. This broke the sequence and left a fraction of the satellite zone unmapped. Mistakes in hour angle and declination settings disqualified a total of three photographs. Two more were fogged in varying degrees by moonlight, whilst one showed no stellar images. After his initial inspection Professor Pickering passed the remainder to Miss L. M. Wells for a more careful examination:

Each plate was scrutinized with a magnifying glass systematically, so that no portion should be overlooked. Each suspicious object was re-examined with a magnifying-glass of higher power, which would show the difference between an ordinary defect and a photographic trail. The remaining objects, about fifty in number, were designated by their approximate co-ordinates on the plates. These right ascensions and declinations were next determined by laying each plate in turn on a Dürchmusterung chart, which is on the same scale of two centimetres to a degree, and making the corresponding images coincide. The number of suspicious objects was finally reduced to twelve. No two of these objects are very near together, although some of them are so marked that they could not be missed, and the greater portion of them should appear on several plates. As the fainter images might be overlooked, all the other plates were examined in which each of these regions was duplicated. In six cases a suspicious object was detected on one of the plates, although absent from all the other plates.[22]

Two plates showed a similar object:

Date	Exposure Time	Position of Object	
1888 GMAT		RA	Dec
Jan 28d 12h 07m	3m	9h 27m·3	+13°.1
12h 26m	4m	9h 28m·5	+12°.9

Though the difference in position corresponded with that expected of a satellite, no trace of the object could be found on three other plates which covered the same area. Moreover, its distance from the Moon indicated that it could not be retained by that body as a satellite.

An object most nearly resembling a satellite was located on a four-minute exposure at 11h 31m, in RA 8h 22m·8, Dec +23°·9. Another suspect featured in a three-minute exposure at 10h 47m, but was not identified on seven other plates of the region. Further a microscope examination 'showed pretty clearly that the suspected object was a defect in the plate.'[23] This explanation also applied to other objects.

'This examination,' Pickering concluded, 'seems to prove that no satellite exists bright enough to have impressed itself photographically upon these plates.'[24] He further concluded:

The greatest diameter a satellite might have, and yet have escaped detection by these photographs, may be inferred from the following considerations: The adjustment of the telescope described above prevented any star from leaving an impression unless it was bright enough to do this in about twenty seconds. Accordingly, stars below the tenth magnitude do not appear on any of the plates, and those fainter than the ninth magnitude appear only on the best plates. The apparent motion of the Moon on the plate being about one-third of that of the stars, the apparent brightness of a satellite would be increased by about one magnitude. We may accordingly conclude that an object of the tenth magnitude would not probably escape detection. Combining this result with the conclusion given above, it is probable that the Moon has no satellite more than 200 metres in diameter. A larger satellite might exist if it was very dark in color, and therefore faint, if it was in the shadow of the Earth during the eclipse, or if its motion relative to the Moon was equal, and opposite to that of the Moon with reference to the stars. In this case it would give the same trail as a star. If the same result is attained in another eclipse, the probability of the existence of such a satellite will be greatly diminished.[25]

Pickering thought, 'A repetition of this experiment ... is therefore much to be desired.'[26] Preparations were in fact made for similar observations during the lunar eclipse of July 28 1888, but had to be abandoned because of cloud and inclement weather.[27] No further attempts were made from Harvard, but in later years William Pickering wrote many papers on the subject of small natural satellites of the Earth, and no doubt possibly re-considered his brother's earlier interest in lunar satellites, but on this point no certain source exists.[28]

In 1895 E. E. Barnard published some interesting comments on the 1888 photographs:

In speaking of the Harvard College photographs of the total lunar eclipses of 1888, I have before me now a glass copy of one of these made during totality January 28, 1888. This picture, though it shows the Moon well in the shadow, does not show the details distinctly; they are more or less blurred and lost through a lack of careful guiding. From the star trails it would appear that the telescope had been adjusted to the motion of the Moon and then left to take care of itself during the exposure.[29]

BARNARD 1895

Barnard himself showed interest in the question during his Lick Observatory days. Like the elder Pickering, he knew of no reason to suppose the existence of a satellite, but thought the idea sufficiently attractive to follow through. He wrote in *The Astrophysical Journal* for December 1895:

> If our Moon had a small satellite revolving about it, such, on account of the enormous brightness of the Moon itself, might never be seen by any of the visual methods. To successfully photograph it under the ordinary conditions would be perhaps impossible because of the spreading of the Moon's light.
>
> If, however, we could obscure the Moon so that it could not illuminate our atmosphere in its direction, we might give a sufficiently long exposure to show any such satellite if it existed near the Moon, and of a brightness so great as the 10th or 12th magnitude.
>
> Such an opportunity is presented during a total lunar eclipse, at which time the faintness of the Moon and its red color would prevent its light spreading on the plate or illuminating our atmosphere. If during any part of this time the satellite should be outside of the shadow and fully illuminated it might be easily photographed.[30]

With this end in view Barnard decided to make a series of photographs with the 6-inch Willard portrait lens of the total phases of the lunar eclipses of 1895, March 10 and September 3, from the Lick Observatory.

The results of March 10 were spoilt by haze, although the photographs showed the Moon clearly in the shadow.

Those obtained from the September search were excellent. The sky was perfectly clear and the duration of totality unusually long. Barnard made six photographs of the total phase on this occasion. To compensate for the motion of the Moon, and in order to secure maximum clarity of photographic image, he adopted manual guidance of the telescope independent of the regular clock drive. At first, Barnard experienced difficulty in selecting a lunar marking sufficiently small yet distinct enough to maintain accurate guiding. Eventually he decided on the Mare Crisium as the most suitable mark, and he kept this carefully and precisely bisected by the wires in the finder telescope, but 'it required constant attention as the motion of the Moon was considerable.'[31] He attended to this task with his characteristic thoroughness:

> That this . . . is shown by the sharpness of the resulting images. I have certainly never seen such exquisite pictures of the Moon as those made during totality with the Willard lens. The details of the surface are clearly and beautifully shown and the Moon stands out from the sky like a beautiful globe.[32]

None of his photographs revealed anything that could be taken for a lunar satellite, and Barnard declared:

> Inasmuch as none of these photographs made during these different eclipses has shown any evidence of a lunar satellite, I think we are fairly justified in assuming that such a body does not exist of sufficient brightness to be detected with our most sensitive photographic plates, and a further search for it therefore appears quite unnecessary.[33]

The survey commenced at 9h 11m, and ended at 10h 51m, Standard Pacific Time. Since Barnard manually guided the telescope on the Moon, the search favoured the greatest elongation zone, where satellite motion would be in the line of sight. Other zones were thus affected by heavy dilution loss due to appreciable trailing of images. His best photograph was taken between 9h 57m and 10h 06m, and represented the Moon whilst still deeply immersed in the shadow. Exposure times ranged from 5 to 23 minutes, and the plates were identical with those used in the Harvard series (8 × 10 inches; sensitometer No 27 made by the M. A. Seed Company of St Louis). As the brightness threshold was two magnitudes below that of 1888, the search had the capability of detecting a satellite of about 200 feet in diameter.

THE IDEA REVIVED

Scientific curiosity alone directed Pickering and Barnard to question the possible existence of a lunar satellite. And it is true to say that to them it was merely an experiment in photographic technique. With no more than an idea to work on, how could it be anything else? Thus, when Barnard concluded that in his opinion further searches were quite unnecessary, his might well have been the last word on the subject. But he had no way of knowing that future technology would demand the assurance of at least one more survey before it was practical to say the Moon has no satellite.

The last search for a satellite of the Moon took place in November 1956, and formed part of a large project to investigate the content of Earth satellite space. It was made by a team from the New Mexico State University, Las Cruces, led by Clyde W. Tombaugh (now professor of astronomy in the University), who as a young student astronomer at the Lowell Observatory secured himself a place in history alongside the elder Herschel, Adams, Leverrier and Galle by his discovery of the legendary ninth planet Pluto in 1930.[34]

The background to the search is both interesting and relevant, and is here briefly described.

Professor T. J. J. See had noted in 1909 when discussing the dynamical theory of satellite capture:

> The possible satellites moving about the Earth, other than the Moon, hardly requires discussion; but as the closed space beyond the Moon is very ample, it is by no means improbable that a small body may yet be found there.[35]

We may also add that a satellite could exist within the orbit of the Moon.

As with most things, that which is on the threshold is usually ignored. It is too familiar, and ostensibly known. To a man astronomers took for granted that Earth has but one Moon. And whilst this was so, if interrogated on the matter none could truthfully assert the correctness of this belief. Few had really bothered to check on the possible existence of tiny satellites:

> Studies of both asteroids and meteorites show that the smaller the bodies the more numerous they are. In whichever case, probability indicates a greater likelihood of the existence of tiny

satellites than of larger ones. This means that the fainter the magnitude limit of a search, the better the prospect of finding such satellites.[36]

Certainly no major institute had accepted the challenge. Satellite space enveloped Earth like some vast uncharted ocean; unknown and largely unexplored.

Of course, Frederic Petit, Director of the Toulouse Observatory, had excitedly reported to the French Academy of Sciences in October 1846, that three of his fellow countrymen had apparently discovered a second Earth Moon in March of that same year. But enthusiasm blinded him to reality, and it is proper to relate his belief to an observation of a particularly conspicuous fireball or bolide.[37]

Several German amateurs tried to substantiate the idea at the turn of the present century, and conducted a number of telescopic sweeps; whilst Professor William H. Pickering advocated an intensive search in 1923. Surprisingly, he renounced the efficacy of the photographic plate as a means of discovering such bodies, simply because he thought they would move too fast to activate one. He obviously forgot that the converse is equally true. A bright satellite would register, but since its trail might span a plate from edge to edge it would naturally be indistinguishable from the trail left by a meteor, and so be disregarded.[38]

In general, the methods suggested by these observers were loosely based on existing search techniques, but as these were geared in the main to the detection of slow-moving objects, they stood little chance of effecting the discovery of Earth satellites whose periods of revolution would range from 24 hours at a geocentric distance of 26,180 miles and 2 hours at 5,000 miles. Such high angular velocities require a new concept in search operations 'drastically different from the technique used in searching for trans-Neptunian planets,' as Tombaugh observed in 1959.[39]

Tombaugh had first been attracted to the question of tracking hypothetical bodies with high angular velocities during his probe of deep planet space. This had commenced in 1929, and culminated in January of the succeeding year with the detection of

Pluto, although the event was not made public until March. The search also incorporated a hunt for what Tombaugh called 'super-asteroids', hypothetical bodies supposed to inhabit the region between Saturn and Neptune. To be visible such bodies would need to be large. The probe had a magnitude limit of 17 and extended to a heliocentric distance of 400 astronomical units. Needless to say, the presumed bodies did not materialise, and the search terminated in 1943.

Now, two years after the discovery of Pluto, the Belgian astronomer E. J. Delporte had accidentally registered the trail of what appeared to be an unassuming minor planet on one of his routine photographs. Acknowledged as new, it was eventually named Amor. Examination of the elements of its orbit, however, disclosed that it was far from ordinary, but belongs to that select band of asteroids whose paths occasionally bring them perilously close to the Earth.

All these bodies were registered photographically, despite trailing loss produced by their rapid motion, as their circuit brought them close to Earth for a few days. This fact left a strong impression on Tombaugh, and induced him to contrast the technique of his search for slow-moving trans-Saturnian objects, with that necessary to reveal hypothetical bodies with high angular speeds. Certain observations in 1943, which had the characteristics of the latter, stimulated further thought on the matter, and directly originated the search for minor Earth and Moon satellites.

In that year Tombaugh and Henry Giclas were at the Lowell Observatory endeavouring to recover the minor planet Adonis as it swung near to Earth. Discovered in 1936, and barely a mile in diameter, this fragment has the same uncomfortable distinction as Amor, and passes about a million miles from the orbits of Venus, Earth and Mars. Naturally it was expected to be faint and difficult to find. For this purpose, we need hardly remark, photography long ago supplanted the visual method. The appropriate portion of the heavens is photographed by a large equatorially mounted camera, the drive mechanism of which is regulated to keep pace with the stars. On the developed plate the latter emerge as dots or point images, whereas the minor planet shows as a short streak or trail, because it is in motion throughout the exposure. But the expected motion of Adonis persuaded Tombaugh and Giclas to dispense with normal practice and adjust the drive of their instrument to follow the planet instead of the stellar background. Unfortunately its position was but poorly known, hence a large area of sky had to be searched. This obliged the observers constantly to reset the micrometer wire of the guide telescope on a selected star during the half hour needed to record the anticipated magnitude.

Adonis was lost, but the idea of an Earth satellite search evolved. Not directly, nor yet in so many words, but in the sense that the mechanics of the recovery attempt provided Tombaugh with the incentive to consider in depth the problem of 'tracking hypothetical bodies at much greater speeds.'[40] Against a background of rocket technology a concept gradually began to take shape.

Rocketry had made substantial progress since the end of World War II, and current developments encouraged the belief that within the next generation space exploration would be elevated from the pages of science fiction. Inevitably, new disciplines would arise and new instruments forged to implement them. Already the artificial satellite was under discussion.

Tombaugh was in charge of optical instrumentation at the White Sands Proving Ground when the two-stage combination of a German V2 and Wac Corporal, Project Bumper as it was officially known, attained a record height of 244 miles in the early afternoon of February 24 1949. Man had taken his first real step into space. Rockets would ascend to greater heights, install satellites in orbit around the Earth, and venture to the Moon and beyond.

The importance of the event underlined the need to ascertain what hazards awaited these missiles outside the confines of the atmosphere—especially of the possible incidence of solid bodies. The shattered condition of the lunar surface mutely testified to the efficacy of the meteoroid. That, however, was only a partial assessment. No clear idea then existed of the character of the first artificial satellites. Everyone expected them to be small, passive objects, mere projectiles that would be tracked by radar, and perhaps by optical means. All very well. What if Earth was attended by a system of tiny Moons? Would not these interfere with radar contact by intercepting transmissions intended for the artificial body, and so confuse the record back on Earth? In the absence of definite knowledge about space immediately around the Earth, the possibility could not be ignored. In the circumstances, Professor Tombaugh felt it needed investigation. The astronomical background was exactly that to which he had been attracted after the unsuccessful hunt for Adonis back in 1943, and which had continued to exert pressure ever since.

By 1952, his discussion consolidated, Tombaugh had worked out 'a plan for searching out any existing small satellites of the Earth.'[41] Essentially it was one of basic research, and was of interest to science in general. Its purpose being to explore Earth satellite space almost to its limit (about a million miles geocentric distance) for solid bodies capable of detection with the available instruments. Tombaugh also envisaged that it would 'provide experience and technique useful in following extremely high altitude rockets and obtaining co-ordinate data along points of their trajectories'.[42]

The search programme was based on the following precept:

In order to detect a small natural satellite, it must be illuminated by the rays of the sun. The large angular diameter of the Earth's shadow would limit the observing periods to the early evening and the pre-dawn hours. High angular driving rates must be used in the direction of the expected motion of a satellite, with a particular rate employed for a given distance, according to Kepler's laws of planetary motion plus the effects of the observer's parallax. Consequently, the search in a plane must be divided up into over one hundred concentric 'depth' or 'distance' zones.[43]

Satellite brightness offered no major problems, as Tombaugh observed in his project proposal. Exceedingly faint astronomical subjects had been photographed before, subjects in all probability much fainter than any extant Earth satellite. But in each case, the motion of the object concerned deviated little from the standard sidereal drive of the instrument used. Even so, the 'high, geocentric angular velocities impose an extraordinary problem of keeping track of the photographed and the un-photographed sectors when the observations are interrupted by the Earth's shadow, daylight, and clouds.'[44] The proposal stated that ordnance missiles had been photographed in daylight to magnitude 3, despite their high speed. The satellite search was to descend to magnitude 13, something like 10,000 times fainter.

To meet these requirements, an equatorially mounted Schmidt or some other fast wide-field camera was to be driven at a rate consistent with that of the hypothetical satellites. Images of the latter would appear as points or short trails, whilst the stellar background would dissolve into a blur of drawn-out streaks, subdued by trailing loss to such an extent that only the brightest stars would show. Large numbers of exposures would obviously be necessary. Star trails on plates exposed for near satellites would go from edge to edge, but that same background in exposures for distant satellites would consist of shorter trails, necessitating very precise tracking.

To avoid the distraction of spurious images, Tombaugh decided on the use of the 'declination offset' principle. In this, an exposure is deliberately broken into two unequal parts, or nearly so, by a short sharp movement in declination near midexposure. This eliminates flaws, provides a means of confirmation, and enables an immediate determination of the object's distance on the basis of circular motion.[45]

Tombaugh finalised his plan in June 1952 and submitted a twenty-five-page project proposal to the Office of Ordnance Research, US Army, for consideration. Eventually it was reviewed, approved and passed for financing by that department. And in December 1953, from the Lowell Observatory, Flagstaff, Arizona, the invasion of Earth satellite space commenced. That, however, is another story. As in his scan of the periphery of the planetary domain, Tombaugh built in another project; a search for possible tiny bodies in orbit around the Moon.

TOMBAUGH AT FLAGSTAFF 1956

The following account of the final search in this lunar trilogy is based on Professor Tombaugh's published report and papers, details of which appear in the bibliography.

Thick haze aborted the initial attempt planned by Tombaugh and Claude F. Knuckles for the total lunar eclipse of January 18–19 1954. However, conditions could not have been better when Tombaugh and two other members of the earth satellite team, Bradford A. Smith and Charles F. Capen, Jr, of the New Mexico State University, Las Cruces, made the next effort on November 17–18 1956, from the Lowell Observatory, Flagstaff. First, the eclipse occurred near the zenith, thus minimising the effects of atmospheric turbulence, light absorption and refraction. Second, the length of totality, one hour and nineteen minutes, was almost the maximum possible. Thirdly, the Moon was only three days from perigee, bringing it some 18,000 miles nearer than at its apogee point, giving a light gain of approximately 12 per cent—circumstances not to recur in combination for several years in the western part of the United States.

The Lowell Observatory had placed three of their most suitable instruments at the disposal of the project; the 13-inch Lawrence Lowell photographic telescope (f5·2), the one used by Tombaugh to discover Pluto; the 5-inch Cogshall camera (f4·5), carried underneath the former by the same mounting, and the $8\frac{1}{2}$–$12\frac{1}{2}$-inch Harold Lower f1·67 Schmidt, the principal tool of the earth satellite survey. Limiting magnitudes of these instruments were $17\frac{1}{2}$, $15\frac{1}{2}$ and 14 respectively.

The expedition went up to Flagstaff on November 13. Since Tombaugh intended to secure point images of the entire range of hypothetical satellites that could orbit the Moon, much of the time prior to the eclipse was spent in rehearsal and practice guiding on bright lunar formations, as this method favours the elongation sectors either side of the Moon, where a satellite has the same apparent motion as the Moon or nearly so, and moreover, if outside the shadow cone of the Earth, is in full phase illumination. Some anxiety was felt in that during totality, surface features might be too faint to allow accurate guidance even on the brightest spots. Preparation of the 5-inch and 13-inch cameras mainly consisted of checking plate speeds and curvature for field focus of the large 13-inch plates, and the intricate reloading device of the plate holders. Because of his familiarity with the instrument, Capen was assigned to conduct the observations with the Schmidt, housed in a dome nearby. Though equatorially mounted, this lacked a regular sidereal drive and guide telescope, and for the Earth satellite search, had been driven by two independant Graham variable transmission mechanisms, one attached to the polar axis, the other to the declination axis. For the eclipse studies these were restored to their former place.

To measure the brightness of the sky around the eclipsed Moon, Smith had constructed a very sensitive sky photometer with equipment from the university, and had fixed it to a portable 5-inch rich-field reflecting telescope owned by Tombaugh. A moonless sky sample was obtained before dawn, after the gibbous Moon had set, one morning shortly before the eclipse, and the apparatus calibrated for saturation exposure against Eastman 103a-o test plates taken with the 5-inch and 13-inch cameras of the Praesepe cluster.

With final adjustments made, drills repeated and the equipment overhauled for smooth operation, everything was ready for the important day. But what was the promise of weather?

Dull, overcast skies heralded the evening of the eclipse. With storm warnings in the background it looked as if the 1954 episode was to be duplicated. Would the cloud break to allow the observations to proceed? It is well known that there is a tendency for cloud to disperse about midnight when the Moon is full and close to the zenith, and at that season Flagstaff has an even chance of clear skies, hence expectation ran high.

Hopefully the instruments were set and the plate holders loaded. Smith was detailed to guide the 5-inch and 13-inch cameras through the 7-inch guide telescope, and Tombaugh undertook to reload the plate holders, since he was most familiar with their intricacies. With the Graham drive reinstated, Capen set the Schmidt to run as near the diurnal rate of the Moon as possible. Still the outlook was bleak. Already the penumbral phase of the eclipse was well under way, but the implacable cloud remained. Just as the Moon's west limb was about to pass into the umbral shadow, the cloud broke and rapidly dispersed enabling the observations to be conducted in perfect conditions. Even the air was slightly mild, and the observers were able to handle the plates and film with bare hands, despite the time of year.

The brightness of an eclipse cannot be predicted. Normally the disk of the Moon assumes a coppery hue, caused by the refraction of sunlight as it filters through the Earth's atmosphere into the shadow cone. Only the long wavelengths survive the intense atmospheric absorption during the oblique traverse, and so the Moon takes on the traditional eclipse hue. On this occasion, however, heavy cloud along the base of the shadow almost completely blocked off the reddish sunlight giving rise to an unusually dark eclipse. In fact it was nearly too dark. At mid-eclipse the Moon was so faint that Smith experienced considerable difficulty in guiding the second exposure on the normally bright crater Proclus, near the east limb. Consequently, for the third and fourth exposures he was obliged to switch to Aristarchus, bright up in the north-western part of the disk, and by then clearing the darkest part of the umbra. Actually, this darkness aided the observations. The maximum practical exposure that may be given with safety at such a time is about five minutes, and for this the observers were prepared. But as the last segment of sunlit Moon edged into the shadow, and the sky grew dark, the photometer was uncovered and directed to a point in the sky a few degrees away from the Moon. This observation informed the observers that instead of the expected five minutes, the 5-inch and 13-inch plates could be exposed for up to fifteen minutes, making it possible then to probe the elongation sectors for possible satellites one stellar magnitude fainter than might have been the case if the eclipse had turned out to be a bright one.

The 13-inch camera was best suited to the search. Far from being a disadvantage, its longer and slower focal ratio enhanced its capability, since the differential drift rate of possible satellites at various luni-centric distances, and in various orbital poses, was less than that of the Moon through the star field (the latter equal to about half a degree per hour). This meant that the camera worked to the advantage of the search in that its plates could be exposed for longer intervals before the sky background fog limit was reached. Effectively this led to a corresponding gain in stellar magnitude.

Three fifteen-minute exposures were taken with the 13-inch (see Plates, pages 49–50), plate size being 14 × 17 inches, the Moon being centrally positioned on each, and the diameter of its image measuring 0·6 inches; four 8 × 10-inch plates with the Cogshall camera, three of fifteen minutes, the other of nine minutes exposure time; and eighteen films with the Schmidt. Because of its fast focal ratio the longest exposure was limited to two minutes. All the plates and film turned out well.

For point images, the 13-inch plates had an effective magnitude depth of $17\frac{1}{2}$, capable of registering satellites 15 feet in diameter at the distance of the Moon, on the assumption of a 7 per cent albedo. If the eclipse had been bright, exposure time would have been cut to barely five minutes with a corresponding loss in detectability potential (25-feet satellites). Careful guiding reduced dilution loss and kept the light of possible satellites concentrated to point images. Each of these plates had an angular scale of 12 × 15 degrees, and portrayed an area of 24,000 miles, but a magnitude loss, incurred near their edges, diminished absolute coverage to about 20,000 miles luni-centric distance. Unfortunately, the greater scale of the plates left the periphery of the satellite environment untouched. Further, as the observations favoured the narrow elongation zones (six degrees out of the whole 360° of the orbital circumference), only a small percentage of the environment was explored to the limit of the instrument.

The Cogshall plates, however, made up for the deficit. They embraced a wider field, namely twenty degrees, the entire region in which the Moon can retain satellites. For point images the magnitude limit was $15\frac{1}{2}$, which would permit the detection of 7 per cent albedo satellites down to 40 feet in diameter. Magnitude dilution due to image trailing was minimal for these plates. With reference to this factor as evident in other regions, Tombaugh wrote in his 1959 report:

> When one computes and plots the trailing losses, each magnitude contour sector takes on a form resembling that of the German Iron Cross; the lesser luni-centric distances are not as wide because of more rapid revolutionary motion. The sectors in front of and behind the moon are the worst because the motion would be broadside to our view and would amount to 4 and 5 magnitudes loss for the 5 inch and 13 inch instruments, respectively. Thus, the magnitude limits are 11 and 12 respectively. Fortunately, these heavy magnitude losses do not extend very far to each side of the penumbral shadow, and they decrease beyond 15 000 and 20 000 miles, respectively. The next worse sectors occupy over half the volume of satellite space, in which the limit of detectable satellites is magnitude 12 and 13 for the 5 inch and 13 inch plates, respectively. It is interesting to note that the 13 inch plates were one magnitude ahead of the 5 inch plates, in spite of the 13 inch scale being nearly 3 times greater and hence three times more sensitive to trailing. It should be remembered that trailing is essentially one-dimensional, whereas aperture area increases as the square.[46]

The small scale and fast focal ratio of the Schmidt entailed little loss from trailing, perhaps less than one magnitude, and Tombaugh reckoned it could pick up satellites of the fourteenth magnitude as point images. Unlike his colleagues, Capen did not guide on the Moon, but swung the Schmidt east and west of it, extending the angular coverage of its $13\frac{1}{2}$-degree diameter film frame.

While still at Flagstaff, Capen carefully examined one of the 13-inch plates under a blink microscope, and Tombaugh blinked most of the other two plates taken with that instrument, besides practically the whole of one pair of the Cogshall series with the

Blink-Microscope-Comparator. This revealed over 200 satellite suspects, increased to 500 after a complete examination of all the plates and film. Most of these, however, turned out to be dust specks and plate defects. Of the rest which took the form of short trails and point images, none survived a rigorous plate-to-plate comparison. Inevitably they were attributed to the debris encountered when searching astronomical photographs for objects at or near the threshold limit of the instrument used to take them.

Except for the relatively small zones before and aft of the Moon, Tombaugh and his team explored the lunar satellite environment down to the thirteenth magnitude, capable of recording bodies 100 feet in diameter, presuming an albedo of 7 per cent. Greatest elongation zones were probed to progressively fainter limits for bodies of from 100 to 15 feet in diameter. Compared to the limits of the earlier searches, 600 feet (Harvard, 1888) and 200 feet (Lick, 1895), this was an exhaustive and definite sweep. Even so, Tombaugh anticipated that some benefit might be derived from the use of large fast-focal-ratio Schmidt cameras in future searches, in order to eliminate possible satellites in the less favourable regions where image trailing produces massive magnitude dilution. He did hope to carry out a further investigation at the next accessible total lunar eclipse in 1960, and made a request to this effect to another observatory for the use of more powerful equipment, but they lacked enthusiasm, and the idea was shelved. Such a search as Tombaugh concluded in his *Quarterly Status Report* No 5, '. . . would have been most interesting . . . for lunar satellites down to five feet in size, as the probability of finding objects would have been much higher'.[47]

Whether this marks the end of the search for a satellite of the Moon only the future will decide. The possibility is unchanged by the failure of these searches. In the meantime, Man obliged where Nature declined, and satellites beyond the credence of Barnard and Pickering have orbited the Moon. But if Tombaugh re-assured the space technologist that he could expect no danger from unknown satellites, his search had a less obvious significance. In the first instance it converted hearsay into fact and completed the historical concept of the Moon, and in the second, it was perhaps the last optical project in the traditional manner to be connected with the Moon. Ranger, Surveyor, Orbiter, Apollo and Luna subsequently became the popular symbols of lunar exploration, not the telescope and the dedicated visual observer. Admittedly the search for ephemeral phenomena on the surface of the Moon gave the latter one last chance of a major contribution to knowledge, but this was more a marriage of convenience with the Apollo programme than anything else. Thereafter the idea of selenography, as previously understood, was diversified through many other disciplines.

2 The Himalayas of Venus

These supposed mountains, then, are high enough to protrude into the upper air to an altitude of over forty miles above the cloud canopy of the planet.
Ellen M. Clerke (c 1893)

As a star Venus is incomparable. It beguiles the eye, enriches the imagination, and conditions us to expect appearances more striking than are evident in any other planet. Yet as an object of telescopic scrutiny it tests the most experienced observer. Confounded by a brilliance that 'dazzles the sight and exaggerates every imperfection of the telescope', to quote Sir John Herschel,

> ... we see clearly that its surface is not mottled over with permanent spots like the Moon; we notice in it neither mountains nor shadows, but a uniform brightness, in which sometimes we may indeed fancy, or perhaps more than fancy, brighter or obscurer portions, but can seldom or never rest fully satisfied of the fact.[1]

Observed by daylight, when its overall brilliance is reduced, and calm conditions prevail, Venus is a different subject. For then the practised eye is quick to discern a pattern of faint markings, less distinct than those of Mercury or Mars, and different in look, but none the less just as real. Especially is this the case when the planet is less than half illuminated, according to our view.

It is at this phase that phenomena of another order begin to emerge. For given steady air, and a good telescope, we become aware of disfigurement in the profile of the planet. It may be truncation of one or other of the cusps, or subtle inflections in their contour, even a ragged, perhaps asymmetric look to the terminator, or a projection at the limb. It may be something more

48

Page 49   Star-field around totally eclipsed Moon taken with the
Lowell Observatory 5-inch Cogshall Camera (attached to the 13-
inch Lawrence Lowell telescope) by C. W. Tombaugh and B. A.
Smith, 1956 November 17–18, at the Lowell Observatory, Flagstaff. Exposure 15 min. The first of the plates taken simultaneously
with the two instruments. The photograph covers the entire region in which the Moon can hold a satellite. The Pleiades are
shown a few degrees above the Moon. Stars trailed because instrument was guided on one of the brightest lunar craters visible
during eclipse. Same scale as original plate.

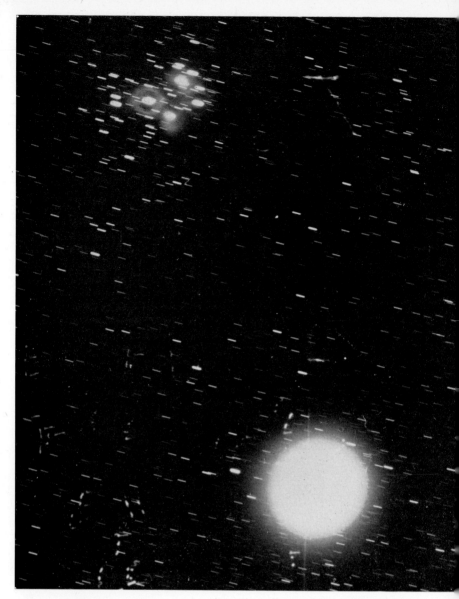

Page 50 Portion of star-field around totally eclipsed Moon taken with the Lowell 13-inch telescope by C. W. Tombaugh and B. A. Smith, 1956 November 17–18, at Flagstaff. Exposure 15min. Plate No 3 of the series. Guided with difficulty on a bright crater during totality. Stars have longer trails because the scale of the 13-inch is three times larger, and trailing three times more sensitive. Photograph is on the same scale as the original plate, and is a small portion of the 14 × 17-inch negative. Check marks indicate the positions of possible satellite suspects, none of which were confirmed by the other plates in the series. Pleiades above the Moon. In spite of the very dark disk of the Moon, the 15min exposure over-exposed the lunar surface pattern.

exotic. A minute luminous point isolated from the bright part of the disk, reminding us in part, as it did the early telescopists, of precisely similar appearances in the Moon, where at the terminator the highest peaks catch the first and last solar rays, to glisten star-like above slopes deeply involved in shadow. To these effects, real or imagined, is attached a curious history. For they are the outward traces of the fabled mountains of Venus. Those prodigious structures which for 250 years excited the same fascination as did the *canali* of Mars, and much the same passion.

THE 'MOUNTAINS' OBSERVED

First described and figured by Francesco Fontana of Naples in 1643, and confirmed shortly after by Briga, as we are told by Webb, these alleged mountains were reaffirmed by Phillipe La Hire, the French astronomer-artist, in August 1700. Observing the planet by day, near to its inferior conjunction he saw '. . . greater inequalities in the termination of light in Venus, than in the moon.'[2] From which the savants of the Paris Academy 'concluded that planet to have higher mountains.'[3] They were next sighted by Bianchini in 1726, and extensively surveyed and measured by the German astronomer Johann Hieronymus Schröter (1745–1816), public official of Lilienthal, near Bremen, during the period 1789–93.

Schröter began his observations of Venus in 1779 with a 3-foot achromatic. Until 1793, when he completed his memoir on its diurnal rotation, he utilised every opportunity to examine the planet '. . . having, in the latter years, observed this planet not only daily, but, as far as the weather and her position admitted, almost hourly through the whole day and evening'.[4] Without such persistence his '. . . trouble for so many years would have been fruitless, as was the case with other observers; *for, in almost innumerable observations, the same thing happened to me as . . . I perceived neither spots, nor any other remarkable appearance, except the unusually quick decrease of light toward the boundary of illumination, which itself was not sharply defined.*'[5]

His luck changed on December 28 1789. Having undertaken his usual survey with a 7-foot Herschelian telescope, armed with powers of 161, 210 and 370, he was astonished to find at 5 pm, that whereas the northern cusp was sharp and pointed, the southern, on the contrary, had a truncated or blunted aspect. More remarkable—slightly separated from it, and enlightening the darkness beyond, he discovered a small luminous speck quivering uncertainly in the chill air (Fig 1). Schröter again saw this configuration on January 31 1790.[6] Two years later, at 7 hours on

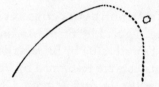

Fig *1* Detached point of light off the south cusp of Venus seen by J. H. Schröter, December 28 1789. The enlightened mountain. South uppermost.

the morning of December 25 1791, it appeared, with the exception of the tiny brilliant, which, however, came into sight at 9h 50m. On the mornings of the 27th and the 30th he satisfied himself of its reality, and convinced that deception had no part in its formation, Schröter supposed this luminous speck to be a lofty mountain whose summit caught the Sun's light before it fell on the surrounding region.[7]

Effects attributable to mountains were further observed in the spring of 1790. Observational conditions in March and April were excellent. Skies were clear and unusually serene. Venus, as he tells us, '. . . was then in Aries at 7° and 8°N declination' and was so high 'that, notwithstanding its approaching inferior conjunction on the 18th of March, at 4 p.m. I could still view it on the 16th, and should certainly have observed it during the conjunction, had not the weather become hazy on the 17th.'[8] Here we quote from his record:

On the 9th of March, 1790, immediately after sun-set, and till 6h 45m, I saw Venus with a seven-feet reflector, magnifying 74, 95, and 161 times, very distinctly, and uncommonly splendid. The southern cusp did not appear precisely of its usual circular form, but rather . . ., inflected in the shape of a hook beyond the luminous semicircle into the dark hemisphere of the planet.

On the following evening, the air being as calm and serene as the preceding one, I observed the planet from 6h to 6h 40m, but on account of some necessary alterations in the apparatus of the seven-feet reflector, I could use only the four-feet instrument, with powers magnifying 134 and 70 times. The southern cusp had its luminous prolongation, but not quite so distinct as the preceding night; but what was more remarkable, each cusp, but chiefly the northern one, had now most evidently a faint tapering prolongation, of a bluish grey cast, which, gradually fading, extended along the dark hemisphere, so that the luminous part of the limb was considerably more than a semicircle.

On the next night, being the 11th of March, when the seven-feet reflector was again fit for use, I found Venus before sun-set, with a power magnifying 95 times. At six o'clock, I saw distinctly the southern point terminating in a luminous streak; which now, as in the evening of the 9th, was longer and narrower than the bright termination of the northern cusp.

The very next, or the fourth evening, gave me a favourable opportunity for this purpose. . . . At six o'clock, the atmosphere being uncommonly clear, I looked at Venus with the seven-feet reflector, magnifying 95 and 74 times. It appeared very distinct, and I ascertained, beyond the possibility of doubt, that the southern cusp projected somewhat into the dark hemisphere, and that from the point of the northern one, the very faint narrow streak of pale bluish light, intermittent in intensity on account of its faintness, but yet permanent as to duration, extended several degrees along the limb of the dark hemisphere of the planet.[9]

Continued observation on March 23 and 24 confirmed the hooked and irregular form of the southern cusp, but on the 25th it was not so perceptible, and by the 30th had disappeared.

Schröter was careful in his analysis of these observations to distinguish between the faint bluish glimmer so conspicuous at the northern cusp, and less so at the southern, and the bright inflection in the latter. His interpretation of the former anticipated that of H. N. Russell in 1898, as the following extracts demonstrate:

> As there can now remain no doubt of the appearance of the pale ash-coloured streak of light . . . extending along the limb of the dark hemisphere of Venus; and as this planet cannot be said, like the moon, to receive some light upon its dark side from our earth, or any other heavenly body, it follows that this light must either proceed immediately from the sun, which, as I have frequently observed concerning the high mountains Leibnitz and Doerfel in the moon (See *Selen. Frag. Chap. 75 and Tab. IV, fig. 6 and 8*), throws its rays directly on the tops of very lofty ridges of mountains; or else that it is a light which partly illuminates the atmosphere of Venus, and partly, being reflected by this atmosphere, marks out by a faint glimmer the limb of the dark hemisphere of the planet, in the same manner as our morning and evening twilight acts upon ours.[10]

And further:

> This light does not appear, as on the mountains Leibnitz and Doerfel in the moon, in single, detached and distant points; but as a continued streak of light, proceeding from the extremities of the cusps. . . . Were this the light of the illuminated summits of a chain of mountains, it would not appear so even, regularly connected, and spherical, as we behold it. . . . Every circumstance, therefore, seems to evince that this phaenomenon is occasioned by a light reflected by the atmosphere of Venus into the dark hemisphere of the planet, being in some measure the light of the atmosphere itself, when illuminated by the rays of the sun; or, in fact, a real twilight.[11]

Against any charge that he had observed just a twilight effect, Schröter wrote in explanation of the bright southern prolongation:

> . . . when the southern cusp extended, not in the true spherical curve of the limb of the planet, but in a somewhat hooked direction, into the dark hemisphere, the pale bluish ash-coloured streak appeared only at the point of the northern cusp, from whence it proceeded, in a true spherical curve, along the dark limb of the planet. On the 10th of March, on the other hand, when the southern cusp did not penetrate so far into the dark hemisphere, the pale streak was perceived at both points, though

somewhat more sensibly at the northern than the southern; and such also were the appearances after the inferior conjunction. These appearances will be thus explained by the effects of a twilight. The bright prolongation of the southern cusp, as it was seen on the 10th and 12th of March . . . must be ascribed to the solar light illuminating a high ridge of mountains situated at this region, whence it appears why this prolongation might not be strictly spherical. The twilight that must have existed at this part, would naturally be eclipsed by the much greater brightness of the light immediately derived from the sun, in the same manner as, on our earth, mountains that face the rising or setting sun, are known to darken the twilight that ought faintly to illuminate the regions situated immediately behind them. Were this not the case, there is no doubt but that a true spherical arch of the limb of the dark hemisphere would have appeared faintly illuminated: and such did we see was the effect of this twilight produced at both cusps when the bright prolongation was not considerable . . .[12]

And of the manner in which these hypothetical mountains affected the appearance of the southern cusp, especially with respect to its blunted form, Schröter further remarked:

Considering the immense height of the mountains, and the great inequalities on the surface of Venus, it is natural to suppose that, as at the times of its greatest elongations, one cusp frequently appears pointed and the other blunt, owing to the shadow of some mountain darkening the extremity of the latter, the same appearance may often take place in the falcated phase of the planet. But the cusp whose extremity is covered by a shadow, will, in this case, so far from appearing blunt, always exhibit a pointed appearance . . . the shadows of mountains, will, no doubt, at times occasion an uneven, ragged appearance, but cannot materially affect the very faint light of the whole.[13]

In this Schröter admitted nothing unusual. Evidently these recurrent irregularities in the contour of the southern cusp were due to some permanent physical cause. If indeed the surface of Venus were uniformly level, he argued, then it should exhibit symmetry in every phase. The horns of its crescent sharp and tapered, dwindling away to mere points, the terminator elliptical and of regular outline. As observation contravened this ideal, and since the irregularities bore a superficial resemblance to those seen in the Moon, it required little further effort to conclude that in all probability they were the visible manifestations of a region of huge elevations proportionately greater in height than any on the Moon, due regard being had to the difference in distance and respective size. A reasonable hypothesis in the circumstances. From the analogy that exists between the Earth and the Moon, it was natural for Schröter to anticipate some inequality of surface. Equally natural for him to suppose, as did William Herschel in 1793, that such '. . . may be, for what we can say to the contrary, very considerable.'[14] Thus at the limb the highest peaks would be seen in profile, whilst those of lesser height would become vicariously visible at the terminator, where they would masquerade as depressions or elevations, according to the angles under which they were respectively illuminated by the Sun, and seen from the Earth. The determining factor being the diurnal rotation of the planet.

Schröter had used his observations of the detached brilliant to determine the rotation period of Venus, and unconsciously influenced by terrestrial analogy, we suspect, had concluded it was 23h 20m 59s·04 in length. Thus a fraction shorter than the terrestrial day. But he carried the analogy a step further. From his observations of the twilight arc at the cusps, in the spring of 1790, he inferred '. . . that the twilight on Venus is nearly equal in its extent to that on our globe.'[15] He fully appreciated that:

. . . the atmosphere of Venus rises like ours, far above the highest mountains: and although we ascribe to that atmosphere the greatest possible transparency, it will still remain a more opake covering than, according to my Selenotopographical Observations, that of the moon appears to be.[16]

Even so, he continued:

. . . it appears that the perpendicular height of the inferior and more dense part of the atmosphere of Venus, which has the power of reflecting the solar light to such a degree, as, under favourable circumstances, to be visible on our globe, where, with a good telescope, it assumes the appearance of a faint ash-coloured light, measures 2526 toises, or 15156 Paris feet.[17]

On this basis, the cloud-filled part of the atmosphere was only three to four miles in height, and this presumed fact not only completed the analogy with the Earth, but set the stage for the final act in Schröter's mountain extravaganza. Communicating his results to the Royal Society in the spring of 1792, Schröter concluded:

> Thus we see a remarkable coincidence in every respect; and yet, though we cannot suppose a smaller, but rather a greater force of gravity on the surface of Venus than on our globe, nature seems, however, to have raised on the former such great inequalities, and mountains of such enormous height, as to exceed 4, 5, and even 6 times the perpendicular elevation of Cimboraco, the highest of our mountains.[18]

The French physicist and astronomer Charles Marie de la Condamine (1701–74), who in 1735 accompanied Godin and Bouguer to Peru and measured an arc of meridian some 3 degrees long near the equator, and spent eleven years in South America, accorded a height of 3,200 French toises to Cimboraco (now Chimborazo). Lalande apportioned 830 toises to one English mile, hence the mountains of which Schröter spoke were in excess of twenty-three miles high! The enlightened peak of December 1789 had been measured and found to be 18,000 toises, or rather less than twenty-two miles in altitude. Whilst the highest peak Schröter supposed to be 18,900 toises in height, again something like twenty-three miles elevation.

Now if Schröter digested this intelligence, William Herschel found it quite unpalatable. 'These extraordinary relations; equally wonderful,' as he sarcastically added, led him to question 'by what accident I came to overlook mountains in this planet, which are said to be *"of such enormous height, as to exceed . . . . Cimboraco, the highest of our mountains!"*'[19] Assuredly it was not through '. . . want of attention, nor a deficiency of instruments,' that 'could occasion my not perceiving *these mountains of more than 23 miles in height; this jagged border of Venus.*'[20]

In short, Herschel disbelieved Schröter on almost every point brought out in his paper, and decided that such extravagance must be admonished by investigation.

Herschel had begun his observations of Venus in 1777 with his 7-foot telescope (two years sooner than Schröter), chiefly to test the contradictory results of Cassini and Bianchini with respect to the rotation of the planet. No progress was made. Detail emerged for the first time on June 19 1780, in the form of a bluish, dusky patch along the terminator, together with an adjoining bright zone. Other spots of various forms appeared on different occasions up to July 3 of that year. Motion was suspected in a spot seen May 21 1783, but it was too indefinite for him to be certain of the fact. This spot reappeared a month later, but generally his experience was the same as that of everyone else.

If Herschel shared the common frustration, it must also be observed that until the publication of Schröter's paper in 1792, his interest had been cursory and erratic. Somehow the incredible findings of the latter compelled him to more closely examine Venus. Even then, intense disbelief rather than a true scientific spirit motivated his interest.

It is not surprising that his observations of 1793 were totally negative. Though he meticulously searched the glowing disk of the planet with his superior optical power, namely his 7-foot and 10-foot reflectors, not once did he suspect any of the appearances recorded by Schröter. Exceptionally, May 13 brought a small success, '. . . the points of the horns appear more blunt than they were last night, and not drawn out to so slender a point.' But he denied the reality of this effect and concluded '. . . this is evidently a deception, owing to the indifference of the night; for great sharpness, and distinct vision, are wanting in every other object I am looking at.'[21]

Though he scanned the rim of the planet for the *enlightened mountain*, and the great ridge in that same region, his experience was always the same. Daily his record read, 'No mountains visible,' 'Not the least appearance of mountains,' 'I perceive no mountains' with such monotony that he might well have had the phrase cut into a die and stamped. Finally on May 12 1793, having again entered a negative report into his journal, he felt impelled to remark:

The slender part of the crescent appears often knotty, but this is evidently a deception arising from undulations in the air; for, with proper attention, the knots may be perceived to change place. Little scratches in the great, or small speculum, may also occasion seeming irregularities; but, with proper attention, all such deceptions may be easily detected.[22]

Things were as he had suspected. No mountains, nothing of the curious phenomena so fully accounted by the astronomer of Lilienthal. His innate suspicions confirmed, and equipped by his experience, Herschel proceeded to demolish Schröter's hypothesis.

Published in the *Philosophical Transactions* (1793), his paper, which included a full description of his observations of Venus from 1777 to 1793, sharply criticised Schröter and rejected his discoveries. Herschel found that his measures and calculations relative to the twilight arc at the cusps proved conclusively that the atmosphere of Venus is much more extensive than Schröter believed, and he attributed the latter's result to defective measures and inferior computation. He also questioned the state of Schröter's 7-foot telescope (which Herschel had made):

... as probably the mirror ... which was a very excellent one, was by that time considerably tarnished, and had lost much of the light necessary to shew the extent of the cusps in their full brilliancy.[23]

He further disputed Schröter's methods, as also his so-called *projection table*, which Herschel hinted had been copied from one of his devices (lamp, disk and periphery micrometers).[24]

'As to the mountains in Venus,' Herschel declared, 'I may venture to say that no eye, which is not considerably better than mine, or assisted by much better instruments, will ever get a sight of them.'[25] However, he did admit that such could exist, and that they might be of considerable elevation. But it was his firm opinion that the cusps were always pointed, and the terminator free of any irregularity. Doubtless Venus presented this aspect on the days when he observed it, and in this connection his records are invaluable; but this is not to say that it was so on other days.

For among his observations, 'but few,' Schröter later emphasised, Herschel *'cannot shew a single one in which he observed at the same time with me.'*[26] 'This is an example to add to many others,' François Arago the Paris astronomer remarked, 'to show the fallibility of negative proofs, even when they come from such a man as Herschel.' He further stated, with regard to the paper as a whole, that it was, 'une critique fort vive, et, en, apparence du moins, quelque peu passionnée.' Perplexed by its unusual severity of tone, Schröter attempted to reconcile it 'to the friendly sentiments which the author has always hitherto expressed towards me, and which I hold extremely precious.'[27] Yet in deference to Herschel, it must be said, misapprehension, not injustice, generated his heated assault.

Schröter effectively clarified his position in his calm and dignified rejoinder, lodged with the Royal Society in 1794, and published in the succeeding year. Herschel, he declared, 'considers it as an equally wonderful relation, that I have SEEN in Venus, in the same manner as in the moon, mountains and shadows of mountains, . . . and that I thence pretended to have determined the rotation of this planet; on the contrary, he considers this last as hitherto undetermined, *because* HE *has never found a trace of mountains.*' But, as Schröter continued:

> I should indeed be surprised that the celebrated author had not, in all the time since 1777, perceived any inequality in the boundary of light, or other appearance of that kind, tending to confirm the existence of very high mountains according to the old observations, were it not that his bold spirit of investigation has been chiefly employed in making much more extensive discoveries in the far distant regions of the heavens, where he has gathered unfading laurels. In fact, the observations which he has communicated from his journals are *much too few* to prove a negative against old and recent astronomers.[28]

And with respect to his own earlier studies, Schröter added:

> *. . . but in other observations, almost innumerable, which I made partly before I had paid any particular regard to the inequality of the horns, and partly in the intervals, I did not perceive, any more than the author,*

*either spots or any thing appertaining to the matter in question; and
consequently our corresponding observations perfectly agree together.*[29]

At this point in the debate, Schröter underlined the flaw in Herschel's critique, by stating:

I have myself also never actually SEEN MOUNTAINS in Venus
AS IN THE MOON, but only *deduced* their existence and height
from the observed appearances. It is even impossible to see them,
according to what I have expressly asserted in my paper on the
Twilight of Venus.[30]

He supplemented this by indicating how the early observers had
examined the Moon without once detecting a trace of the great
cordilleras at its south pole, to which Schröter appended the
names of Leibnitz and Doerfel. 'And yet these high mountains
are really there.' Further, had not Herschel perceived projections
in the ring of Saturn, and by them determined its period of
rotation? From so many accordant observations of Venus, therefore, Schröter failed to see how his conclusion, namely that its
mountains '*bear nearly the same proportion in height to her diameter, as
those of the moon do to the diameter of the moon,*' should be found so
disagreeable, 'especially since all my observations hitherto, as for
instance those on the visible luminous spots in the dark part of
the moon, on the apparent changes of the moon's surface, etc.,
have been confirmed by others.'[31]

Coincidental with Herschel, Schröter had re-observed Venus
during the spring and early summer months of 1793. Against
twenty-five observations adduced by the former, Schröter opposed
one hundred, taken at various times on different dates. On four
occasions they observed the planet simultaneously, significantly
to agree on what they saw. A fact to which Schröter affixed some
importance. Throughout this period he again discerned definite
traces of the great southern cordilleras, and once more the faint
glimmer of the enlightened mountain off the southern cusp,
besides the anomalous contours in the cusps themselves.[32]

The following extract from his record is typical of the majority,
and is here given because it was made in the presence of Ernst

Florens Friedrich Chladni (1756–1827), who was the first to prove that meteorites are not of terrestrial origin:

> March 9th, 6h 15m. p.m. (1793). Venus being near her greatest elongation, both horns appeared pretty pointed, with a power of 250, and a fine soft image; they were also *both alike*, but with the slight difference, that close to the southern horn *a very minute particle projected, which seemed to be rather separated from the rest of the enlightened part.* At 8h 2m, *the air being clear, a projecting inequality shewed itself with certainty at the southern horn.* . . . . It was found the same with 288 of the 13-feet. As our own atmosphere was then very clear, that of Venus also seemed to be purer than usual; for with both reflectors, and particularly with the 13-feet, Dr. Chladni, *as well as myself,* enjoyed a magnificent view of the arch of illumination, which seldom presents itself so well to the eye; *the image being uncommonly clear and distinct. To both of us the boundary of illumination, toward which the light became very dim, appeared* (be it ever so much contradicted) *not only nebulous, and not sharply terminated, though sensibly sharper than usual, but also very evidently unequal and rugged, with faint shades between,* as I have often seen it, but never so plainly. In truth, the appearance, as each declared, *was very like* the image of the moon at the time of her quadratures, only that the boundary of light was sensibly less sharp, and the faint shadows between were not almost black, but in some measure like the dark spots of the moon's surface, grey, yet darker than the other parts.[33]

If such inequalities were solely illusions, or the product of the planet's atmosphere, Schröter argued, was it not reasonable to expect them more often? How was it they were chiefly disposed at the southern cusp and along the terminator, and most noticeable about the time of greatest elongation, 'when the eye looks perpendicularly through the dense atmosphere of Venus . . . and by no means in the small crescent'? If these 'small indentations and darker places' were truly atmospheric effects, similar phenomena should invest the enlightened part of the disk. And if they were deceptions, he continued, why did the southern cusp invariably appear the smaller? How was it, when the planet was observed through light cloud, that this same cusp always disappeared sooner than its broader counterpart? Unless he was very much mistaken then, these appearances, as authenticated by his latest observations, were only explicable on the assumption 'that the planet Venus has very considerable mountains and elevated ridges; and indeed the most and the highest in her southern hemisphere.'[34] Subdued by Herschel's heated attack, Schröter discreetly avoided further mention of his measurements of the heights of these formations, being content to describe them in general terms. In concluding his brief, he challenged the future by stating:

> If any astronomer shall think it worth the trouble to observe *Venus, not barely now and then, at whatever time of the day it may be, but continually, with the same persevering zeal, and when the weather is favourable almost hourly, about the time of her greatest distance from the sun,* I am convinced that he will certainly perceive the rare phaenomenon in question, just as well as I have done. If, contrary to all reasons which hitherto appear, I should hereafter be convinced that I was deceived, I would myself, willingly and impartially, bring the offering to truth; and so much the more readily, as no indirect views have ever led me on, but I have been actuated solely by an irresistible impulse to observe; and because I certainly shall never have reason to be ashamed of the observations I have laid before the world, which have always conducted me to new truths.[35]

Thus Johann Hieronymus Schröter, astronomer extraordinary of Lilienthal, convinced himself and the world that Venus is a rugged and mountainous planet. The great debate had ended partially in his favour. And yet it was, strictly speaking, a remarkable triumph. Schröter was a persevering and enthusiastic observer, selflessly devoted to his interest. Single-handed, under adverse conditions, he virtually founded what came to be known as comparative lunar and planetary study, but, strangely enough, he was constantly handicapped by a strong overplay of imagination, and an apparent inability to subordinate theory to observation—shortcomings which seriously impaired the value of his results. It was this tendency to rash and often fanciful conclusion to which Herschel objected, and which prompted him to castigate Schröter in such acrimonious terms. Perhaps not justified by the

actual circumstances, these strictures were by no means un-
warranted by the facts as Herschel saw them. Yet even though he
could not authenticate his details of precise elevation (the Lilienthal micrometers were wholly unsuited to this delicate task[36]),
Schröter had completely vindicated his incredible mountains, or
at least the effects of their existence. By the end of the nineteenth
century they had acquired factual status and were acknowledged
in the works of Webb (who wrote of 'the mountainous surface'),
Flammarion, Guillemin, Proctor and other astronomical *littera-
teurs* of the day. Some early science fiction even embodied the
idea, and portrayed a planet dramatic in content. Such confidence
had its origin in numerous confirmatory observations. In effect,
Schröter had not been mistaken. The accuracy of his record was
beyond question, even though he had perhaps fallen down in his
physical interpretation.

Between 1833 and 1836, the great German astronomer J. H.
Mädler confirmed truncation and other anomalies in the southern
cusp, as also in the terminator. He was followed by De Vico, at
Rome, in the period 1840–1, and by Gruithuisen in 1847. During
the winter of 1853–4, James Breen, the Cambridge astronomer,
examined Venus with the Northumberland refractor and frequently remarked on the apparent bluntness of the southern cusp,
in contrast to the relatively sharp, tapered aspect of the northern.[37]
He also described unusual curvature in the terminator. Earlier
terminator indentation had been remarked by Flaugergues in
1796; by Valz and Lamont, of Munich. Whilst Fritsch, in 1799
and once more in 1801, with a small telescope, saw very much the
same appearance. Arcimis, the Spanish astronomer, observed a
prominent notch in the southern cusp in 1876. Professor F. Porro,
Turin Observatory, with an 11-inch Merz refractor, saw the
terminator humped in the centre, with a depression on either side,
May 14 1892.[38] W. Alexander, 8½-inch refractor, reported a
similar feature, possibly the same, during the early part of June.[39]

A curious configuration was witnessed by De Vico and Palomba
of the Roman College Observatory, in April and May 1841.
Using a 6¼-inch Cauchoix refractor, they noticed near the northern

cusp, when the crescent was slender, an oblong dusky spot, subsequently involved in an oval luminosity, which they compared to an oblique view of a lunar crater. On successive nights they followed its progress as it drifted towards the terminator, until finally, half the ring vanished into the nocturnal hemisphere, at which stage the spot formed a dark notch in the terminator between two bright arms. According to their description it seemed to resemble a deep valley encircled by lofty mountains. It had a diameter of 4·5 seconds of arc. Webb states in his *Celestial Objects*, that Gruithuisen and Pastorff observed like appearances. William Lassell, when in Malta in January 1862, with his 4-foot mirror, figured a comparable detail.[40]

When Venus was at its greatest brilliancy in the evening sky of 1881, W. F. Denning discovered a dusky indentation not far from the northern cusp, described as '... extremely small, and looks like a crater, though I could not be certain of this.' Confirmed on the next night, March 31, and well seen April 5, by which time it had shifted away from the cusp, but this apparent motion could be explained by the movement of the planet relative to the observer, and the increasing length of its cusp.[41] Denning had seen a similar apparition in the same quarter March 29 1873.[42]

Modern times also have their share of facsimiles. F. Sargent, of Bristol, England, discovered a '... most pronounced notch', just below the northern cusp at 4 pm, on February 15 1913 with his 10½-inch reflector, and held it steadily until 5.30 pm.[43] It was independently found by H. McEwen, of Glasgow, Director of the Mercury and Venus Section of the British Astronomical Association, shortly after 5 pm, the same day with a 5-inch refractor. He described it as '... a notch or indentation ... one fourth of the planet's diameter from the north cusp.'[44] February 22, McEwen wrote '... the appearance of the notch was considerably altered. ... It no longer presented the form of an indentation; all that was visible was a sharp kink-like bend in the terminator.'[45] At the next observation, March 2, the feature had disappeared, and the terminator presented its usual uniformly shaded and regular outline. We may add that another prominent indentation was seen in this area December 21 1954. The evidence is flimsy, but is it possible that the observations by Denning, Lassell, McEwen, and Sargent all relate to the configuration seen by De Vico in 1841? If so what is the cause? Was Schröter basically correct?

A well marked hemispherical depression was reported in the terminator below the southern cusp by several American and English observers during the first week of March 1953. Clearly seen on the evening of March 10 by the writer, it seemed bounded on its northern and southern flanks by dim projections and was preceded on the disk by an oval bright spot. It received the nickname Bartlett's Cirque, from its resemblance to a lunar crater, and after its discoverer Dr James C. Bartlett, Jr, of Baltimore. W. J. Wilson and the writer were witness to a comparable appearance, but in the opposite phase, December 17 1970 at 11hrs UT. Conditions were not good, yet the feature was distinctly seen with the former's 8½-inch reflector. It had the aspect, as with the others, of a shadow-filled lunar crater at the terminator. More recently, throughout the 1972 eastern elongation, another English observer, W. J. Beetles, perceived traces of deformities in the southern half of the terminator with his 6-inch reflector. Enough has thus been given to indicate that since Schröter's time, his accuracy has been fully and precisely verified, and the reality of these effects as objective features is now accepted without question.

It must be understood, however, and this is appended more as a caution than anything else, that reports of irregularities in the narrow crescent must ever be treated with reserve. We have quoted what Herschel had to say on this point with regard to his observation of May 12 1793. To balance the record, and to show that he also appreciated the danger, let us quote Schröter on this question of doubt:

> It is scarcely necessary to put the reader in mind, that small, undulating, knotty inequalities of the boundary of light, in such observations, must not be taken for true inequalities, or mountains of Venus. In general, these small crescents, as the enlightened part lies obliquely to the eye, are not well suited for observing the true inequalities of the boundary line, or any spots

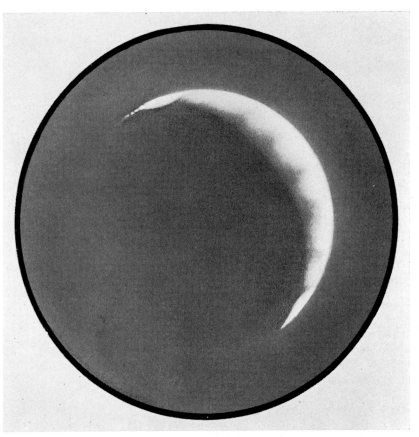

Page 67 Irregularities at the south cusp of Venus. Observed and drawn by the writer 1959 September 30 06h 40m to 07h UT, with a 4·5-inch equatorial × 186. The broken aspect at the south possibly due to unsteady air, perhaps explaining the nature of the high peak reported by Schröter in this area.

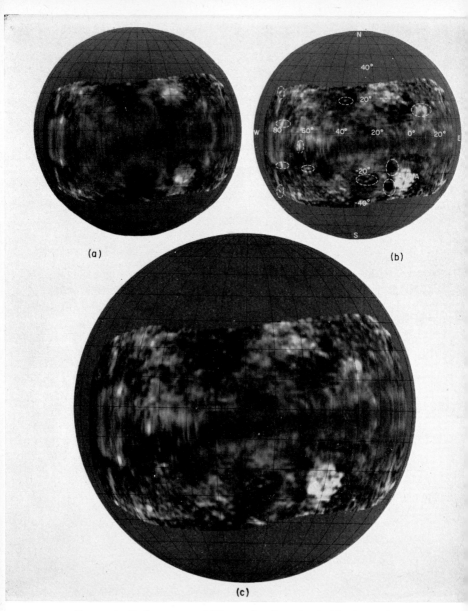

Page 68 Radar map of Venus by R. M. Goldstein and H. Rumsey Jr, based on studies on seventeen dates around inferior conjunction 1969 with the 85ft dish and the 210ft dish, of the Goldstone Tracking Station, California.
(a) Cumulative radar map from all data but with no regard to the north-south ambiguity.
(b) Co-ordinate frame for Venus. Zero meridian selected to pass through feature Alpha.
(c) Result of removing most of the north-south ambiguity by the least-squares technique.

there may happen to be. For such observations, we should be assiduous in attending to the planet, about the time of its greatest distance from the sun.[46]

Perhaps we may remark that the reference to the boundary of light relates to the terminator, and the greatest distance from the Sun relates to the greatest angular distance as observed from Earth, hence greatest elongation.

POLAR HOODS AND STAR SPOTS

So far in our review of observations confirmatory of Schröter, we have studiously avoided all mention of the southern cusp itself apart, that is, from its recurrent truncation. Here we are embroiled with a somewhat different class of irregularity, which owes its recognition only partly to the industrious observer from Lilienthal. For Schröter, despite his numerous observations of the regions, failed to identify specifically the great bright clouds at either cusp. Superficially, they compare with the bright deposits that mark the polar areas of Mars and Earth, but of their true cause we have no knowledge. Leaving aside any suggestion of comparison with the polar spots of the latter bodies, for the coincidence is purely optical, we may remark that E. M. Antoniadi firmly opposed the idea that the bright clouds or hoods at the cusps of Venus were objective formations, and attributed them to a function of phase.[47] William Herschel hinted at such an explanation in 1793, when he wrote of the limb brightness, thus:

> With regard to the cause of this appearance (limb brightness), I believe that I may venture to ascribe it to the atmosphere of Venus, which, like our own, is probably replete with matter that reflects and refracts light copiously in all directions. Therefore on the border, where we have an oblique view of it, there will of consequence be an increase of this luminous appearance.[48]

If the bright hoods were in fact constrained in visibility to the crescent phase, there would be some justification in upholding this argument. But the truth is, they are not thus restrained. Reputable visual observers have delineated them in the gibbous phase, although faintly. Certainly they are conspicuous when the planet is less than half illuminated, but in the circumstances this is to be expected, since then the planet is nearer to us. The death knell of the phase-function hypothesis came in 1927 after F. E. Ross had photographed Venus in ultra-violet light with the 60-inch Mt Wilson reflector. These photographs, in showing decided traces of the hoods, completely accredited the older reports, and gave credence to the idea that Venus has, for want of a better term, polar caps. Last century Webb and Phillips suspected they might be due to snow and ice, but this can now be discredited. The chemical composition of the hoods, however, is not here considered. Our purpose is to present a brief account of these features in anticipation of the final chapter in the mountain saga; namely, that of the Great Southern Plateau, or the Great Southern Uplands to which Schröter and Trouvelot subscribed.

It is to Franz von Paula Gruithuisen, Professor of Astronomy, Munich University, that we owe the discovery of the hoods in 1813. He delineated them particularly well on December 29 1813.[49] From their apparent immobility and brilliance he naturally thought they were permanent, and he compared them to the caps of Mars, and inferred in the process that they probably marked the poles of rotation. Vögel and Lohse saw them clearly in 1871 from the Bothkamp Observatory.[50] Holden at the Lick Observatory noted something of the sort December 15 1877. G. V. Schiaparelli extensively studied the southern spot in the interval December 8 1877 to January 6 1878, and further in the summer of 1892. From these observations he concluded Venus rotates on its axis in a period synchronous with that of its period of revolution around the Sun.[51] In recent years the hoods have been the object of close observation by members of the Mercury and Venus Section, British Astronomical Association, and the equivalent group of the Association of Lunar and Planetary Observers (USA), besides various organisations in Europe.

No doubt now exists as to the objectivity of the polar hoods of Venus. In appearance they take on the aspect of bright spots of variable size flanked and bisected, especially in the southern cusp, by deep shadows (see Plate, page 67). It is attractive to suppose they represent the atmospheric poles. The fact that they cover the north and south points, and that nowhere else along the terminator are there persistent bright spots, does suggest these are preferential points.

Usually, but not invariably, the southern spot is the largest, the brightest, and the most distinct. It is, moreover, the most often seen. That at the north is weak, diffuse, and less easy to define, and occasionally absent. Significantly, neither Barnard nor Lowell ever commented on their presence. Drawings by Lowell in the period 1896-7 do show traces, and there is some evidence to suggest that he did see rare polar brightenings, but the evidence is not strong. E. C. Slipher, at the Lowell Observatory, writes that polar brightening was sometimes seen, but not continuously. Always, however, when it was evident it favoured the southern cusp.

A marking upon which there is almost universal agreement, visually and photographically, is a dusky band perpendicular to the terminator and adjacent to the southern cap. Schiaparelli showed it well, as did Trouvelot. Lowell christened it 'a sort of collar round the southern pole.'[52] He described it at length, and spoke of knots and enlargements along its stretch. Quénisset's drawing from his photograph of February 23 1921 equates with this description. Niesten and Stuyvaert, of the Royal Observatory, Brussels, noted it often between 1881 and 1895.[53]

What emerges from this brief discourse on the polar hoods of Venus is simply this: that of the two, the southern by far excels the northern in brilliance and extent. And this is remarkable, for here it was that Schröter discovered the majority of his appearances. Yet it is rather odd that he failed to mention the brightening. No matter—what now concerns us are those final observations affirmatory of his belief that the extreme southern latitudes of the planet are characterised by a prodigious mountain ridge.

Effects analogous to his enlightened mountain, that detached speck riding off the cusp extremity like some faint star, were reported by the Baron Van Ertborn, of Antwerp, who observed the dissevered condition of the cusp on August 13 and 20 1876; by

Lohse and Wigglesworth with the latter's 15-inch refractor, January 2 1886;[54] and by W. F. Denning, who commented:

> ... on certain occasions, when the planet has been ill-defined in passing vapours, it was most easy to believe that a fragment became detached from the extremity of the cusp, just in the manner described by Schröter.[55]

Percival Lowell, referring to his series of 1896–7, wrote '... Schröter's projection near the south cusp was also clearly discernible as well as two others, one in mid-terminator, one near the northern cusp.'[56] On June 9 1948, the American observer T. L. Cragg, noted vague, faintly luminous protrusions at the northern and southern cusps. Of irregular form, they were seen twenty-four hours later to have extended further around the darkened limb of the planet.[57] Only one observation of a fragment broken off the northern cusp has been located. It was made by an English amateur, A. P. Holden, April 2 1870. To him, the '... N. horn appeared divided at its extremity by a dark line, making the extreme point like a bright speck detached from the other portion.'[58]

Over the years Schröter's hypothesis attracted quite a number of disciples. Yet its most ardent supporter came to light in 1892 when he published the results of almost twenty years of observation of Venus with large telescopes in good conditions. His name was Etienne Leopold Trouvelot (1827–95). He had emigrated to the United States from France in 1852, because of some political trouble, and had joined the Harvard College Observatory. Ultimately, he returned to France in 1882, where he worked with Janssen at the Meudon Observatory, Paris.

Trouvelot made two series of observations. The first, which began in the autumn of 1875 and ended early in 1882, was made from Harvard. The second, from Meudon, started in November 1882 and continued to 1892. The first set included 565 observations and 162 drawings, and the second 179 observations and 133 drawings.[59]

His most striking observations were first reported to *The*

Observatory in a letter dated Cambridge, Mass, March 20 1880, from which we extract the following:

> From Nov. 13, 1877, till Feb. 7, 1878, two remarkable white spots, strongly reminding me of those seen on Mars, have been observed on the opposite limbs, near the extremity of the cusps. The southern spot, which always appeared the brightest, became very prominent from Jan. 16, 1878, till Feb. 5, and appeared then to be composed of a multitude of bright peaks forming on its northern border a row of brilliant star-like dots of light. After the inferior conjunction, which occurred a few days later, the white spots were no more visible. This phenomenon has been independantly seen by Mr. Seagrave of Providence, with his 8-inch refractor.[60]

Trouvelot apparently saw this configuration three times in this period. His description of February 2 1878, as Venus drew near inferior conjunction, given here in translation from his memoir, is worthy of preservation:

> The polar spots are distinctly visible, the southern one being the more brilliant. Their surface is irregular and seems like a confused mass of luminous points, separated by comparatively sombre intervening spaces. This surface is undoubtedly very broken and resembles that of a mountainous district studded with numerous peaks, or our polar regions, with numerous ice-needles brilliantly reflecting the sunshine.

Later that same evening:

> The polar spots seem bristling with peaks and needles. This is especially the case in regard to the southern spots, which seem entirely formed of brilliant points. On the north polar spot is a luminous peak which seems to project outside the limb; or if it be not so, the parts surrounding this peak must be dark enough to be confused with the background of the sky. But I am fairly certain that this spot projects outside the limb.[61]

Barnard, Molesworth, and McEwen, at times saw minute bright spots which they thought might be lofty mountain summits, perhaps snow-clad, reaching up through the cloud. Professor Carl Venceslas Zenger, Prague, wrote of a shining speck close to the south cusp on September 30 1876.[62] He later wrote (February 17 1877):

> ... of very high reflecting points on the surface of *Venus* (perhaps snow-covered high peaks) being visible, like on the Moon, before the solar light reaches the lower parts of the surrounding surface.[63]

R. Langdon, a well known English amateur of Trimdon Station, saw April 17 1873 at 8 pm, '... two exceedingly bright spots on the crescent—one close to the terminator, towards the eastern horn, and the other in the centre of the crescent ... appeared like two drops of dew ... glistening in such a manner as to cause the surrounding parts of the bright crescent to appear dull by contrast.'[64] Henry Pratt of Brighton, England, also noted, January 29 1870 at 5h, in good conditions, '... a tooth of light near the S. cusp ... evidently a spot on the terminator higher than the adjacent regions.'[65] He also saw other odd features in the terminator, notably a strong indentation near the presumed equator. W. F. Denning reported a bright spot in the north cap, April 1873.[66] Still another in the same region on March 30 and 31 1881.[67] The Rev T. E. Espin also spoke of tiny brilliants near the cusps as 'detached mountain peaks'.

Trouvelot confirmed that the blunting of the south cusp is almost always seen when the phase is large, and that the limb invariably retains its symmetrical form. He also commented on the frequent irregularity and unusual extension of the cusps, particularly in the south, and drew attention to a flattened S-shape aspect periodically assumed by the terminator. To him the bright cusp caps identified the polar caps, there being hardly any doubt as to their reality. Two more or less bright spots, 180° apart, which maintain that situation do suggest, he argued, a special relationship which conforms to what we would expect if they were true polar zones.

Of the bright star spots, and the other irregularity observed in those areas, again there could be no doubt. They were static and could only be the visual result of high peaks thrusting through the dense cloud mantle; mountains of such altitude as to clear the upper surface of that cloud and glisten in the sunlight. Amongst these peaks Trouvelot fixed the poles of rotation, the axis being inclined 10° to the plane of the orbit. As a gesture in support of the great polar plateau, he cited photographs of Venus taken by Arago and Bouquet de la Grye, in Mexico, during its transit of December 6 1882, which showed a great bulge in the vicinity of the southern cap, indicating that '. . . there exists in the south of the planet a zone characterized by a great elevation between two depressions.'[68] Measures of this bulge gave it an elevation of about sixty-five miles.

REFLECTIONS ON THE HYPOTHESIS

No one who has seen irregularity in the figure of Venus will deny the fact that at best it is an inconsequential affair, still further diminished by reason of the angle under which it is observed. Indistinguishable from the effects of bad seeing and irradiation, it is vulnerable to rejection on these, besides other counts.

Atmospheric turbulence often imparts to the terminator a toothed or jagged look, precisely as figured by Fontana, La Hire and Bianchini, and perhaps this explains the origin of the hypothesis. Schröter avoided implication in this charge by his remarks on the aspect of the slender crescent in unsteady air. However, W. F. Denning was rather doubtful that Schröter had avoided it:

> . . . there is strong negative evidence among modern observations as to the existence of abnormal features; so that the presence of very elevated mountains must be regarded as extremely doubtful, if, indeed, the theory has not to be entirely abandoned. The detached point at the S. horn shown in Schröter's telescope was probably a false appearance due to atmospheric disturbances or instrumental defects. Whenever the seeing is indifferent, this planet assumes some treacherous features which are very apt to deceive the observer, especially if his telescope is faulty. Spurious details are seen, which quite disappear from the sharp images obtained in steadier air with a good glass.[69]

To Denning, as to others, the answer was plain. All the irregularity seen in Venus was false. Rippling of the image instituted the blunted cusp and the 'outlying lucid point', in much the same way, again to quote Denning,

> that passing air-waves and resulting quivers in the image of Saturn's ring will sometimes produce displacements, so that the observer momentarily sees several black divisions, and the edges are multiplied and superimposed one on another. Refraction, exercised by heated vapours in crossing objects, is obviously the source of all this.[70]

Indubitably atmospheric agitation does modify and alter the telescopic image, distorting its form, and converting its detail into a confused blur, even to the extent of total effacement. It is an attractive and winning suggestion, but the effect is normally overall not particular. What is seen at one cusp should coincidentally be visible at the other. And why should the southern cusp be favoured more than the northern? Except on the supposition that it is affected by some basic abnormality? There is also a failure to explain what I. G. Lohse observed on January 2 1886 with the 15-inch refractor at Scarborough, England. He noticed a narrow streak of light, about one-twentieth the diameter of Venus, extending the northern cusp, yet quite separated from it.[71] T. L. Cragg commented on similar prolongations at either cusp on successive days in April 1948, as already mentioned.[57]

Such observations, where the same phenomena were seen successively, cannot be dismissed as attributes of poor seeing. We may conclude, therefore, that although atmospheric conditions can and do produce false appearances, they cannot be held to account for relatively sharp asperities and indentations, and may be less important in this charade than is generally supposed.

On the other hand, contrast effect and irradiation affords a satisfactory rationale. The surface of Venus, that is the apparent surface, is brilliant but not uniformly so. From peak brightness at the limb, its light fades by degrees until at the terminator it is replaced by a hazy duskiness that itself merges imperceptibly with the darkness of the nocturnal side. Local darkenings and brightenings within this zone are not unusual, and often lie contiguous to the terminator. If we have two bright areas in such a position, by contrast the region between will appear depressed. Conversely, a bright spot hemmed in by dusky shadows will be arbitrarily enhanced in light, and will appear to project beyond the curve of the terminator. Dr James C. Bartlett, Jr, Baltimore, made an interesting comment in this respect in 1953:

> . . . if we consult the Ross photos (taken in ultraviolet), we notice many distortions along the bright limb and some at the south cap. These, of course, arise from density differences in different parts of the Venusian atmosphere and to qualitative differences in the clouds, both combining to affect locally the penetration by ultraviolet which has small penetration at best. In this sense, therefore, the Ross projections are not true projections at all but mere artifacts of the technique. The writer proposes that for a somewhat similar reason projections may appear upon the limb during visual observation. If selective absorption alters the wavelength of reflected light, thereby changing its color, an eye perhaps less sensitive to the altered color than to a contiguous area might well see the latter as much brighter and therefore projecting beyond that portion of the limb recognized as duller.[72]

Soft undulations in the terminator which are manifest as long, low swellings and depressions, such as were conspicuous to Alexander,[39] Mascari,[73] Poro,[38] and Zona,[74] in 1892, and to many others before and since, together with unusual curvature of the terminator as a whole, evidenced by Breen,[37] Mädler,[75] and Niesten and Stuyvaert 1881 to 1895,[53] and the curious S-shape configuration frequently espied by Trouvelot,[76] to mention the most notable, are probably thus accounted for. Similarly, projection of the bright cusp caps above the general level of the limb, again most pronounced in the south, has its cause in contrast and irradiative action. Relatively sharp, localised features, for example, the independently seen McEwen–Sargent object of February 1913, fall into a different category. Most probably they represent genuine irregularity.

Optical imperfection of the telescope, and the fact that much of the irregularity reported was seen by users of small telescopes, provide further arguments to negate the idea. Whilst the former is of considerable import, the latter is not strictly correct. Many spectacular effects were discovered by experienced observers at the oculars of powerful telescopes. Statistically the balance favours the small telescope, qualitatively the emphasis is laid by the larger aperture, that is, above 10 inches.

To clarify the situation, let us classify and annotate the irregularities. Three main groups may be distinguished: (i) Cusp phenomena, (ii) Terminator phenomena, and (iii) Star spots.

(i) Cusp phenomena

    (a) Truncation of the cusps. Conspicuous at dichotomy and sometimes during the narrow crescent phase. Favours the southern cusp, and is recurrent. Well attested, and undoubtedly a real effect.

    (b) Irregular projections of the cusps diverting their extremities away from the geometric norm. Not consistent with the Twilight Arc. Again reasonably well seen and confirmed, and probably real.

    (c) Inflections on the inner margin. Often seen and possibly attributable to a compound of unsteady air, irradiation and reality.

    (d) Detached luminous points and fragments hovering off the cusp extremities. Delicate and questionable appearances, but have been steadily held in large glasses in good seeing, and observed successively.

(ii) Terminator phenomena

    (a) Smoothed undulating forms, either long low protrusions, or shallow depressions. Possibly caused by contrast effects and irradiation induced by the uneven intensity of the terminator shading.

    (b) Serrated aspect, founded in inferior seeing.

    (c) Sharply defined projections and indentations, most likely true objects.

    (d) Abnormal curvature of the terminator, transgressing the line of elliptical curve. Possibly due to intensification of the terminator shading, either as darkening or brightening. [77]

(iii) Star spots

    (a) Points of high luminosity chiefly clustered around the cusps, mainly the southern, but visible elsewhere on the disk. Usually seen only with large telescopes.

Clearly, then, few of these appearances can be explained with certainty—though they do seem to have a basis of reality. However, mountains will not explain this rebellion in the geometric perfection of the planet's figure.

In the first place, we are reminded of the enormous height to which these features must rise in order to be visible at the distance of Venus. Witness Schröter's embarrassment on this point. We have noted how most observations adduced as evidence of mountains were made with small telescopes. Further, some were taken when Venus lay near inferior conjunction. Now at this time the planet subtends an apparent angular diameter of about sixty seconds of arc. Its actual diameter is around 7,800 miles. From this it follows that in this position one second of arc corresponds to approximately 130 miles. Visual observations are seldom conducted in other than mediocre conditions. Consequently it would be rather difficult for a small telescope, say of six inches aperture, to reveal a projection, a mountain in profile, unless it attains a height of 0·3 seconds of arc. Such a structure would rise fully 39 miles above the apparent surface. Mountains of this calibre are basically unacceptable. Writing in *Nature*, December 23 1967, Professor Carl Sagan, Smithsonian Astrophysical Observatory, stated that the hydrostatic pressures at the base of such structures would exceed the yield and tensile strength of all abundant geochemical materials.[78] In simple terms, they would flatten under their own weight.

More cogent reasons influence the second objection. Mountains would appear at regular and predictable intervals. With the exception of the south cusp blunting, however, the irregularities are ephemeral and haphazard in their displays. Further there is no real connection between any of the recorded markings. This is a telling and significant negation. What is more, all this presupposes clear, uninterupted views of the actual surface. We know well that the atmosphere of Venus is more extensive than was appreciated by the earlier observers. The globe of Venus rotates unseen beneath a dense canopy of lemon yellow cloud, opaque to visible radiation, with few breaks apparent. Mariner II informed us in 1962 that this cloud stratum extends from a base height of about 45 miles above the surface to a maximum altitude of around 60 miles. All we ever see of Venus then '. . . is the shining upper floor of a firmanent of cloud,' to quote Miss Agnes Clerke, the astronomical historian. Elevation measures, as in the case of Schröter, were referred to the apparent surface. It is not, therefore, simply a question of the height of the visual appearance, but that plus the height of the cloud layer, giving to the enlightened mountain a phenomenal altitude of 83 miles! Because of the continuous and unbroken condition of the cloud the foregoing strictures cannot be relaxed. For this is a fatal objection to the mountain hypothesis, yet it provides us with a possible answer to the riddle, always assuming we are dealing with real appearances.

Without invoking redundant hypotheses in contradiction of what is known of the planet, and without invoking formations that could not exist, we perceive another set of phenomena capable of explaining all the irregularity—namely clouds. These project an acceptable and perfectly reasonable substitute.

Great towers and billows, marking the confrontation of cold and warm currents, surge to immense heights in our immediate experience, and lead us to fancy in their panorama all the characteristics of mountainous terrain. Deep gorges and valleys, dull and shadowed; castellated ramparts of formidable aspect; pinnacles and peaks glistening as if snow-capped—it is as though every conceivable geological formation has its atmospheric counterpart. Permanent in one sense, yet impermanent in detail and form, clouds condense and dissipate; dissipate and re-form in endless pageant. They appear in one place at one time, in another the next; importantly they gather at different heights dependent on a variety of factors, and they cast shadows.

Is it not feasible to imagine a similar state on Venus, where the action of atmosphere is more complex and violent? Thus cloud, in a view reduced by distance, would extend or curtail the limit of vision and by its shadow efface those portions normally visible, giving vent to depressions and projections along the terminator, and unequal appearances in the horns of the crescent: and all this would naturally derive from height variations in the cloud, and temperature differences of the underlying surface condition. We commonly witness, notably at sunrise and sunset, a flare of light on a distant hillside as some highly reflective surface turned by the rotation of the planet to the correct angle catches and briefly reflects the glory of the low Sun. So on Venus. Tiny specks shine out against the dull backcloth of yellowish cloud; no more, we concede, than Sun glint from high cloud tops. Likewise the broken extremities of the cusps. High altitude cloud on which the Sun's light lingers long after vacating the lower scene.

Surface relief has a profound effect on atmospheric circulation, cloud formation and distribution, and on Earth clouds are not infrequently rooted to mountain chains. As the air-flow encounters these obstacles it develops two movements, upwards on the windward side and downwards on the lee. Consequently the locale becomes one of increasing turbulence, in which cloud is destroyed by the descending current, created and thrown up to great heights by that in ascent. Given suitable conditions this leads to the formation of cloud considerably above the mountain tops.

Most probably the surface of Venus is diversified by some inequality, comparable in stature to that of Earth. Perhaps here and there it may be considerable. On this basis we may assume that its effect on atmospheric circulation may be marked out in the clouds by periodic appearances, such as observed at the southern cusp. In that case it may be explained by the prevalence of high altitude cloud occasioned by the presence of some lofty range or plateau. This reaction of the underlying surface on the clouds could satisfactorily account for all the observed phenomena.

We may quote some data in support of this proposition. Since

1942 Venus has been extensively photographed in yellow light by the French investigator, H. Camichel, at the Pic du Midi. His observations have led to the discovery of certain low contrast dusky shadings which disappear at intervals, as if obscured by cloud at a higher level. From this, Camichel inferred that these dusky areas might be quasi-permanent. A. Dollfus, Meudon Observatory, has suggested two explanations. Either they represent indistinct glimpses of the actual surface, or cloud banks associated with surface anomalies.[79]

The latter idea is not inconsistent with the latest radar observations. Microwave radiation is not inhibited like visible radiation. It can penetrate the cloud and make contact with the surface. Statistical analysis of the resultant echoes, which are due to reflection at, or just below the surface of the planet, enables us to estimate surface roughness. As the radar reflectivity of any material is dependent on its electrical characteristics, besides its smoothness, calculation of its average dielectric constant is possible, and in turn this permits us to work out something of its composition, hence that of the scattering medium. Obviously this is a complex and involved process, with uncertainty ever in attendance. Even so, recent developments in observational techniques and the application of more sensitive instrumentation have brushed aside centuries of speculation and have lifted the veil of mystery that has enshrouded Venus.

Most of the energy in a Venus radar pulse resides in its leading edge, indicating strongest reflection from the nearest parts of the planet. If the surface were uniformly level then the amount of energy reflected would follow a definite and predictable pattern. Initial American scans demonstrated that the surface is apparently smoother and more compact than either that of the Moon or the Earth, at least in terms of irregularities greater than a radar wavelength. Whilst calculation of the average dielectric constant has yielded a value consistent with that of quartz, suggesting a rock surface with dry dusty conditions obtaining.

But in 1962, R. M. Goldstein and S. Zohar of the Jet Propulsion Laboratory, California Institute of Technology, using the facilities of the Goldstone Tracking Station, discovered that certain portions of the surface return more energy than expected, marking in consequence areas of high radar reflectivity. These appear to be permanent surface features, as they exhibit a measurable Doppler Shift and rotate with the planet. Expressed in visual terms these radar bright regions imply or rather suggest irregular, broken terrain, or, more probably, enhanced localised roughness. Further studies in 1964, 1966, 1967 and 1969 (see Plates, pages 68, 117) from Goldstone and other stations confirm the original findings, and reveal that these radar bright regions are of complex structure. Now whether they are distinct geological formations ie, craters, mountains, plateaux, boulder-strewn areas, lava flows etc, as some believe, or merely inherent peculiarities of surface, has still to be determined. But the fact that mountains have once more been invoked is interesting.[80]

In this connection we may recall that when the United States spacecraft Mariner II passed Venus on December 4 1962, its on-board infra-red radiometer registered a spot in the southern hemisphere not far removed from one of the radar bright regions, some 20°F colder than its immediate surroundings. This also has been tentatively ascribed to a high surface structure.[81]

Whether these radar and radiometer objects are mountains remains to be seen. The interpretation emanates from dubious premises, and it is possible that future, more refined studies may impose a startlingly different picture to that currently envisaged. It is curious that the fabled Himalayas should continue to dominate the horizons of Cytherean knowledge, against all reason and fact. Extraordinary that they should again be resurrected, unless it is that they really do exist. It is as though the cicada voice of Schröter were chuckling at modern naïveté. And we are obliged to ask: Are we confronted by a romantic survival, or hard fact? At least the latest findings provide a departure point from which to view anew the whole concept.

3 'An Unexplained Observation'

Already bright stars observed by our fathers have disappeared from the maps of the sky.

Camille Flammarion (1894)

WHILE presiding over the Nashville, Tennessee, meeting of the American Association for the Advancement of Science in August 1877, Professor Simon Newcomb (1835–1909) was approached by a young man of twenty, a photographer by occupation but a diligent watcher of the skies in his leisure hours. Having acquired a 5-inch refracting telescope he naturally sought how best to deploy it in the observation of celestial phenomena. Although he presumed the young man could do no more than amuse himself, Newcomb proposed that he undertake to search for new comets, in his opinion the only field open to an observer in his situation.[1]

Much taken by this suggestion, the young man began to sweep the heavens for new comets, and on May 12 1881 found a faint one near α Pegasi, which he saw on two mornings, but never again. A real success rewarded him on September 17, however, when he picked up the comet which brought his name prominently before the astronomical world (1881 VI). This was quickly followed by Comet 1882 III, and in the next year but one by 1884 II. The succeeding year yielded two comets, 1886 one, and 1887 three more. Whilst he discovered every new comet of 1891. In fact, for the ten years after beginning this work he shared with the celebrated W. R. Brooks the great majority of such finds. Five times he won the Warner prize offered for these discoveries.[2]

In 1883 he left the photographic business, having received a fellowship in Astronomy at Vanderbilt University. He was promptly placed in charge of the Observatory attached to the University, and continued his researches with the 6-inch equatorial. That same year he observed the occultation of the well known star β Capricorni, by the Moon. As the Moon passed over the star, the observer noticed that instead of instant disappearance, the process was gradual. Now the interval between diminution, and complete extinction of the star's light took only a few tenths of a second, but it was enough to alert the observer that the matter needed investigation. Further observation of the star with the 6-inch failed to reveal any peculiarity. He called attention to the event in the astronomical journals, and conjectured that the star might be a binary. Subsequently, S. W. Burnham with the $18\frac{1}{2}$-inch equatorial of the Dearborn Observatory, Chicago, confirmed this was so. We may further add, that the Nashville astronomer during that same year re-discovered the *Gegenschein*; a phenomenon first reported in 1854, though without then attracting interest.

Acclaimed for his discoveries and skill as an observer, Edward Emerson Barnard (1857–1923), the young Nashville astronomer, emerged from obscurity to become one of the most famous observers of his day.

BARNARD DISCOVERS JUPITER V

At the time of his death in October 1876, James Lick, the San Francisco millionaire, had already selected the site of a large observatory high up in the Californian Coastal Range, on one of the peaks of Mount Hamilton. Affiliated to the University of California, the observatory became fully operational in June 1888, and came under the directorship of Professor E. S. Holden, late assistant to Newcomb at Washington, DC. Holden intuitively selected young astronomers of promise for positions in the new establishment, and at his graduation Barnard was offered and accepted the post of assistant astronomer. That was in 1888.[3]

During the next year, he began the important series of observations that was to make his care and skill legendary. At first he worked solely with the 12-inch refractor, and continued his photographic studies of the Milky Way. Every now and again, he received permission to observe specific objects with the great 36-inch. And beginning in July 1892 he was officially assigned the regular use of this telescope one night a week, that night being Friday.

Two months later Barnard made his first major discovery, namely the fifth satellite of Jupiter. This happened on the night of September 9. He had just completed an observation of Mars, and had measured the position of its satellites, when around midnight he suspected a diminutive speck of light closely following Jupiter, not far from satellite III, then rapidly approaching transit. Though almost lost in the glare of the planet, Barnard was able to follow it and ultimately divine its true character. To appreciate the exact circumstances of the discovery, we can do no better than quote Barnard himself from *The Astronomical Journal* of October 4 1892:

Among other things that I have devoted the instrument to on my nights, was a search for new objects. Several of the nights have been bad, and have more or less limited the investigations.

Nothing of special importance was encountered until the night of September 9, when, in carefully examining the immediate region of the planet *Jupiter*, I detected an exceedingly small star close to the planet and near the 3d satellite. I at once suspected this to be a new satellite. I at once measured the distance and position-angle of the object with reference to satellite III. I then tried to get measures referred to *Jupiter*, but found that one of the wires had got broken out and the other loosened. Before anything further could be done the object disappeared in the glare about *Jupiter*. Though I was positive the object was a new satellite, I had only the one set of measures, which was hardly proof enough for announcement.

I replaced the wires the next morning. The next night with the great telescope being Professor SCHAERBELE's, he very kindly gave the instrument up to me, and I had the pleasure of verifying the discovery, and secured a good set of measures at elongation. In these observations, and those of the succeeding night, only distances from the following limb of *Jupiter* could be measured. These were observed with the wires set perpendicular to the belts. The planet was thrown outside the field, the satellite bisected, and then the limb brought in and bisected also. This method would not permit any measures from the poles of the planet for latitude. On the 12th, I inserted a strip of mica, carefully smoked, in front of the field-lens, for occulting the planet. This served admirably, permitting the satellite and planet to be both seen at once, and measures from the polar limbs could be made with great ease. The observations of the satellite from the 12th, were all thus made.

And with respect to its brightness,

Just what the magnitude of the satellite is, it is at present quite impossible to tell. Taking into consideration its position, however, in the glare of *Jupiter*, it would, perhaps, not be fainter than the thirteenth magnitude. . . . In general the satellite has been faint—much more difficult than the satellites of *Mars*. On the 13th inst., however, when the air was very clear, it was quite easy.

It is scarcely probable that this satellite will be seen with anything less than 26 inches, and only with that under first-class conditions.[4]

The detection of this body, the first satellite of Jupiter to be found since 1610 when Galileo discovered the four principal moons of the Jovian system, affirmed the vigilance of its author and raised him to still greater heights of fame. It is especially interesting, therefore, to contrast this observation with one Barnard made a month earlier, and which curiously enough to this day is still unexplained.

THE 'STAR'

Until daybreak on August 13 1892, the weekly vigil with the 36-inch had been comparatively uneventful. Barnard had first turned to Mars, making several micrometer measurements of its satellites and scrutinising the face of the planet itself, estimating and measuring some of its more salient albedo features in the process. He had then switched to Jupiter, measuring its angular diameter and watching satellite II (Europa) emerge from occultation; last contact of this event being timed at 04h 02m 29s Pacific Standard Time. Some four minutes later he noted the central meridian passage of the Great Red Spot.[5]

With sunrise less than an hour away, he traversed the great tube through an arc that led directly to Venus, brilliant but low in the south east. After twenty minutes or so absorbed in examining the elusive detail of this planet, then approaching dichotomy, Barnard became aware of a star in the same field of view 'which must have been of at least the 7th magnitude to be seen in strong daylight . . . a half hour before sunrise.'[6] Intrigued by its prominence in such conditions, he made a quick note of the time, o4h 50m PST, and estimated it to be '1' south of Venus and 14s ± preceding'.[7] Barnard, of course, tried to locate it more exactly by micrometer, but the attitude of the telescope tube 'with reference to the high chair made it impossible to get the measures before daylight killed it [the object] out.'[8] Indeed, as he later remarked, the star was 'so low that it was necessary to stand upon the high railing of a tall observing chair,' placing him in the hazardous situation of having 'to hold on to the telescope with both hands to keep from falling.'[9] Needless to say, in that position measurements were quite out of the question.

From his approximate fix with relation to Venus, then in RA 6h 52m 44s, and Dec +17° 11'·8 (in the constellation Gemini, not far from the Milky Way), Barnard deduced the position of the star to be about RA 6h 52m 30s, and Dec +17° 11'·0. Corrected to the epoch of his reference chart (*Bonner Dürchmusterung*. 1855.0), this became RA 6h 50m 21s, and Dec +17° 13'·6.[10] But as Barnard discovered:

> There seems to be no considerable star near this place and the object does not agree with any BD. star.
>
> Unless this was one of the brighter asteroids (not Ceres, Pallas, Juno, or Vesta, however, which were elsewhere) I am unable to account for the observation.
>
> The elongation of Venus from the Sun was about 38° which would exclude the possibility that the object was an Intra-Mercurial planet, but it does not preclude the possibility of its being a planet interior to Venus, though such is not probable.
>
> . . . a reflection from the image of Venus is out of the question. No other star was seen near Venus on this date.[11]

THE OBSERVATION QUESTIONED

For no clear reason, Barnard 'hesitated about calling attention to the observation' until fourteen years later. Writing to the *Astronomische Nachrichten* from Chicago, March 22 1906 (he had transferred to the Yerkes Observatory of the University of Chicago in 1897, and taken a Chair of Astronomy in the University), he stated, '. . . now that it is best to put it on record.'[12] As to be expected, it met with disbelief. One European critic, Rudolf Pirovano, in a brief note to the same journal (July 31 1906), went so far as to suggest that Barnard was confused as to the date of the observation.[13] If so, then the position of the star could be changed to correspond with that of a known star.

In the face of this criticism Barnard remained adamant:

> So far as I can remember, the conditions under which the observations were made are correctly stated in A.N. 4106. They are borne out by the original notes which follow the sketch in my note book.[14]

In fact, his rejoinder of September 26 1906, in which he reconstructed the events of his sojourn inside the dome of the Lick refractor on the night of August 12–13 1892, is rather an excellent example of his outstanding abilities as an observer. He refuted Pirovano's criticism by quoting his observations of Mars and its satellites, together with the timed details of his work on Jupiter and Europa. He also confirmed the time in relation to sunrise on that date:

> In Volume I Publications of the Lick Observatory, p. 296, there is a table for sunrise at Mt. Hamilton which gives for Aug. 13, sunrise at 5h 23m. This is 33 minutes after my observation, and is the 'half hour before sunrise' referred to by me.[14]

Of the way the observation had been conducted, Barnard explained:

> In reply to Rudolph Pirovano (AN. 4117) I would say that with the large telescopes of this country 26° would be considered a low altitude for the observation of objects in certain parts of the

sky. These telescopes are not expected to be used near the horizon. When it is required to observe near the horizon, it is necessary to have a high observing chair in addition to the elevating floor. To one who has not used these great telescopes this fact will perhaps be new.[14]

And further, in regard to the chair actually used:

> There were two observing chairs at the Lick Observatory; one which was always used—the regular observing chair; another— a great cumbersome affair—which was made, I believe, to get at the opening cut in the tube about 10 feet back from the eye end, for getting at the photographic plate and for guiding purposes when the 33 inch photographic correcting lens was on. This heavy chair was seldom used by me, or for any other purpose than that stated above, and was evidently not used on the night in question.[15]

That his 'memory may not have gone too far astray' in this respect, Barnard wrote to the Director of the Lick Observatory, Professor W. W. Campbell, asking him:

> . . . to have the 36-inch set to hour angle—4h 39^m. 9 and declination +17°, with the elevating floor at its highest point, and to give me the distance from the eyepiece to the floor.[15]

In effect, therefore, to reconstruct the circumstances of August 13 1892 when the strange star was seen. Campbell complied and reported:

> With the telescope set at east hour angle 4h—40m and declination +17°, the eye piece of the 36-inch is 11 feet (eleven feet) above the highest position of the floor. The top step of the smaller observing chair is 8 feet above the floor, and the top of the large chair is 15 feet above the floor.[15]

Obviously Barnard had recalled correctly. He had used the smaller chair. Any remaining doubts as to the authenticity of the observation simply evaporated, but an explanation still could not be found.

Professor F. J. M. Stratton, Solar Physics Observatory, Cambridge, England, thought it possible that Barnard had seen a nova or exploding star. Dr Joseph Ashbrook reiterated this view in 1956.[16] Visible by day, it is feasible that the star faded before its position became readily available in the evening. As the object was situated near the dense star fields of the Milky Way in Gemini, which is a most likely source of such objects, this hypothesis has everything to recommend it to serious attention. Dr Ashbrook observed that if a large telescope found a blue variable star of between the 18th and 20th magnitude in the position described by Barnard, then assuredly Barnard's mystery star of August 1892 would have been recovered.

This is not an unknown possibility, as Dr Ashbrook demonstrated. He quoted the strange affair of D'Agelet's star. In 1783, the French astronomer J. D'Agelet, l'Observatoire École Militaire, Paris, was cataloguing stars when in the July he measured the position of a sixth magnitude star in Sagitta. Though well seen on three nights, namely the 26th, 27th and 29th, later observers could not find it. In 1951, Harold Weaver, of the Lick Observatory, was comparing plates taken in red and blue light with the 60-inch Mt Wilson reflector, when he found a 19th magnitude object on one plate almost in the position of D'Agelet's missing star. Uncommonly blue in colour and of variable light, this star has all the characteristics of a nova in its post-maximum phase.[17] In the circumstances it is reasonable to suppose that Barnard had the same experience as his French predecessor. Can we follow Dr Ashbrook in his assumption of Nova Geminorum 1892? If not, then what did Barnard observe?

4 A Strange Celestial Visitor

. . . *a comet seen during the totality of the Eclipse of the Sun of May 17, 1882, which comet was never seen again, and whose history and circumstances will probably remain for ever undisclosed.*

George F. Chambers (1909)

JUST before midnight on Sunday, August 7 1921, the following telegraph message was received at the Harvard College Observatory for distribution to other observatories:

> Star-like object certainly brighter than *Venus* three degrees east, one degree south of Sun seen several minutes before and at sunset by naked eye. Five observers. Set behind low clouds. Unquestionably celestial object. Chances favor nucleus bright comet, less probably nova.[1]

The telegram came from the Lick Observatory, and had been composed by its Director, Professor William Wallace Campbell (1862–1938), and Professor Henry Norris Russell (1877–1957), Director of the Princeton Observatory.

But the alert was too late. As mysteriously as it came the object had vanished. To this day its brief eruption on that sunset sky remains a puzzling enigma. Others saw it at an earlier hour, but so indefinitely that the observations tell us nothing of its character. Like Barnard's object of 1892, it is totally incomprehensible. Strange, too, that it chose to appear in the same month as the latter, and to the same station. An odd circumstance, really, yet it is only coincidence.

Other than Professors Campbell and Russell, and the other

92

members of their party, whom we shall presently name, the observers included H. C. Emmert, a Detroit, Michigan, medical practitioner; four English amateurs, Lieutenant F. C. Nelson Day, Ferndown, Dorset; S. Fellows of Wolverhampton, and two young students John L. and Gilbert Kershaw, Barrow-in-Furness; and, in Europe, several inhabitants of Plauen, Saxony.

The Lick observers were the last to sight the object, yet it was their telegram that first drew attention to its existence, and it is Professor Campbell's description in the October 1921 issue of the *Publications* of the Astronomical Society of the Pacific, that provides us with the most reliable data. Let us reconstruct how Professor Campbell and his companions came to make the discovery, if so it can be called.

'WHAT STAR IS THAT?'

It all began on the date of the telegram. Professor Campbell had been watching an unusually vivid sunset from the verandah of his official residence on Mount Hamilton, together with his wife and four guests, Professor Russell, Captain Eddie V. Rickenbacher, renowned racing driver and World War I flying ace, and Major and Mrs Reed Chambers.

High-altitude clouds streaked with brilliant hues inflamed the western sky. Between them and a shallow stratum of thin clouds that hugged the horizon, the sky was perfectly clear and transparent. As the lower limb of the Sun adjoined the apparent skyline, 'which consisted of haze, smoke and light clouds,' its image

> began to pass thru the interesting series of geometrical figures so frequently observed here, especially in the summer months. Starting with the ellipse, the succeeding figures resembled the old-fashioned Rochester lamp shade, the straw hat now in vogue with men, an extremely elongated ellipse, the cigar form, and finally, at disappearance, the knitting needle form. The last figure endured for at least a minute of time.[2]

It was at this juncture that Major Chambers casually asked, 'What star is that to the left of the Sun?' Captain Rickenbacher commented he had already seen the same star, and had been looking at it for several minutes, 'but had not mentioned it because he supposed it was well known.'[3] Major Chambers retorted in similar vein, adding he too had seen it a minute or two before. Professors Campbell and Russell, and Mrs Chambers spied it immediately and kept it in view while Mrs Campbell hurried indoors to collect a pair of binoculars. Within the minute she returned and handed them to her husband, who snatched a glimpse of the object before it declined from sight behind the cloud horizon. In that moment Professor Campbell opinioned it was probably Mercury. Professor Russell disagreed, and said it was entirely too bright for that planet.

The cloud afforded no further opportunity for study, and the observers settled down to compare notes on what they had each seen. All agreed that it had been starlike. Of its position and colour:

Captain Rickenbacher said it was six solar diameters from the Sun, and that the line joining Sun and object made an angle of 45 degrees with the vertical line thru the Sun's center. Major Chambers indicated graphically the angle, and his estimate agreed with that made by Director Campbell as 50 degrees southerly from the vertical. Russell and Campbell estimated the distance from the Sun's center at three degrees. Campbell thought that the color of the object was yellower than a star in that position would be, but it must be remembered that the Sun at disappearance had a true zenith distance approximating 93 degrees and that the true zenith distance of the object must have been between 91 and 92 degrees. *Venus* has frequently been observed to set below the western mountains at approximately that zenith distance, as a bright object and with color not distinctly orange or red. Russell and Campbell agreed that the object observed was brighter than *Venus* would have been if seen in the same position and circumstances as the strange body.[4]

After dark Professor Russell determined the position of Mercury from the *Nautical Almanac* and confirmed his earlier comment. The planet had been below and to the right of the Sun. Similarly all the other planets were accounted for. 'What could the object be?'

The observers had no reason to doubt that the object was genuinely celestial, as in the binoculars its diameter still seemed stellar. It seemed to partake of the diurnal motion of the stars in that it moved down toward the lower cloud stratum and disappeared behind it. To imagine an object in our own atmosphere which would have supplied all of the phenomena described seemed entirely too difficult.[5]

Its astronomical character established, two hypotheses were suggested:

Its brightness, as seen on the sunset sky, its position close to the Sun and its sudden appearance as an unobserved object, suggested strongly the nucleus of a comet which had approached the Sun in such a direction as to escape notice. There seemed no alternative except that of a new star, such as, but much brighter than, the one which suddenly appeared in June, 1918, in the constellation *Aquila*.[6]

That they had been witness to a strange happening in the heavens could not be denied, but what? Obviously other stations would have to be alerted. Should this be done immediately, or delayed pending confirmation at the next available opportunity? Indecision was momentary; the matter was carefully studied and the directive clear, and at 11 pm, the telegraph clicked the news to Harvard.

The Lick observers waited at sunrise on the Monday, and again at sunset, and yet again at sunrise on Tuesday, the 9th, but the object failed to appear. That same day Professor Campbell drew up a report of all that had transpired on the Sunday evening; the report previously mentioned. As he wrote, two things happened to deepen the mystery. First, his telegram of the 7th was published in Harvard College Observatory *Bulletin* No 757, August 9 1921. Second, the news was telegraphed to the astronomical bureau at Copenhagen, and given coverage in the next issue of the *Astronomische Nachrichten*.[7]

CONFIRMATION

Within forty-eight hours newspapers of both hemispheres informed their readers of the strange new body near the Sun. Leader writers enlarged upon the topic, and scientific journals speculated as to its nature. Yet it was a fragile thing of which they wrote, both in fact and substance. Here and there, people began to search their memories, and some vaguely remembered how, on or about the day concerned, they too had seen a luminous object in a cloud-strewn sky just above the recently set Sun.

Writing to the *English Mechanic and World of Science*, on August 14, one week after the event, an English amateur astronomer, S. Fellows, of Wolverhampton, described what he had seen:

> A few days ago the statement appeared in our newspapers that an observer at the Lick Observatory had seen, on the evening of August 7, a bright star near the Sun, and that it was visible without optical aid. My experiences on that evening may be worth recording. About 8.30 on that Sunday evening I noticed a bit of very clean sky over the place of the recently set Sun, and, thinking I might pick up Jupiter and Saturn, I took my binoculars (power about 3) and commenced to search. I soon alighted upon a bright object, which I at the moment thought was Jupiter, but the next moment I saw was not the planet at all; neither did I think it was a star. It was elongated in the direction of the Sun, and was of a distinct reddish tinge. I judged it to be about 6° from the Sun and a very little south of him. It was unfortunate that I was not able to get the telescope on the object, as it was too low. I only held it for about three minutes, as clouds came and hid it.
>
> If the object which I saw was the same as that seen by the American observer, and if our estimates of its position were fairly accurate, then the object must have passed inwards towards the Sun some 3° in five or six hours. It must also have greatly increased in brightness in that time, as the object which I saw could not be compared to Venus.[8]

Here the object is described as elongated, whereas Professor Campbell says it looked starlike (but he had it in view for only two seconds); apart from this slight difference however, Fellows undoubtedly saw the same apparition.

Almost a month later, another sighting turned up. It had been made by a Lieutenant F. C. Nelson Day and others at Ferndown, Dorset, about 7 pm, GMT, on August 7. It was communicated to *Nature* (September 8 1921) by Colonel E. E. Markwick, the well known amateur astronomer and ex-president of the British Astronomical Association (1912–14). Apparently the Ferndown observers, in a clouded sky, had noticed an object some 4° distant from the Sun in the following direction, which they estimated to have been of stellar magnitude −2. Though conditions were not good, they were absolutely positive of the identification. Their magnitude estimate, it must be added, is irrelevant since the Sun was still above the horizon at the time of the observation.[9]

Whether these reports were based on notes actually made when the object was seen is unknown. Most likely they are the product of memory stimulated to recall by the press stories; Fellows partly admits as much. No doubt though, they do seem to verify the observation from the American astronomical station.

Less certainly, according to a postcard from Dr Max Wolf, Director of the Königstuhl Observatory, received at the Lick Observatory on September 22 1921, and a notice in the *Astronomische Nachrichten*, many inhabitants of Plauen (lat +50° 30'; long 12° 7' E), Saxony, Germany, saw a bright object in the evening sky of August 7, 7h 35m, GMT, in RA 11h 06m·7, and Dec +7° 9'.[10] But this position almost coincides with that of Jupiter (RA 11h 24m, Dec +5°·1). Did the good people of Plauen then, perhaps unacquainted with the current configuration of the planets, confuse their unconscious memory of Jupiter with the new object of which they read in the newspapers?

THE MYSTERY DEEPENED

Evidence exists to suggest the object was in fact observed before August 7, but as this only came to light weeks later, its reliability is under suspicion. Two reports are concerned. One from Barrow-in-Furness, England, the other from Detroit, Michigan, USA, both of the same date, August 6. The former was made before dawn, the latter during the late afternoon.

The first is based on the testimony of two young amateurs who were out looking for Mercury in the pre-dawn light. The observation came to the notice of the late S. B. Gaythorpe, a prominent member of the British Astronomical Association. Having read of the Lick object both in the press and in the Annual Report for 1920–21 of the Comet Section of the aforementioned body, he somehow heard that an identical phenomenon had been observed locally. He located the source and arranged an interview. 'Note on a Luminous Body, 22° West of the Sun, seen before Sunrise on 1921 August 6' dated November 15 1921, and read at the November 1921 meeting of the British Astronomical Association, gives us the result of this contact:

> The object was seen by two young students of astronomy, John L. and Gilbert Kershaw, at about 1921 August 5d 16h 5m G.M.T., [*Greenwich Mean Astronomical Time beginning at noon, discontinued after January 1 1925*] when looking for Mercury, and is described by G.K., who saw it only with the naked eye, as being very similar to the appearance of that planet during its evening apparition in the first week of June. Soon after it was first noticed, the body was lost to sight behind a cloud. Just before it finally disappeared, it was seen for a few seconds by J. L. K. through a small telescope, × 20, in which, however, it did not present any sensible disc. Owing to clouds and the strong twilight its position with respect to the stars could not readily be determined, but it was seen to be a little to the left of a vertical line through Capella, which I compute was at that time in azimuth 106° 30' E. of S., at an altitude of 48° 40', the longitude and latitude of the place of observation being 12m 51s W. and 54° 6'.6 N.[11]

The young observers had estimated the position of the object by alignment with distant landmarks, and the position of the observing site by reference to features in its immediate vicinity. From theodolite observations made on the alleged spot, Gaythorpe reconstructed the circumstances under which the object had been observed. He inferred it had an apparent altitude of about 5° 40', '. . . certainly appreciably less than 5° 58' and its azimuth about 107° 55' E. of S.—being midway between two landmarks bearing 106° 45' and 109° 5', respectively'.[12] Allowing 9' for differential refraction, this placed the object in RA 7h 35m, Dec +14° 55', according to his calculations.

Both these coordinates are inconsistent with the object having been Mercury. The latter was then in azimuth 115° 4' and altitude 7° 50'. In addition to its altitude being considerably greater, the planet—had it been visible—would have been seen 6° to the *left* of the more northerly landmark instead of about 1° to the *right*, as was the luminous body.[13]

Commenting on this observation Dr A. C. D. Crommelin, Director of the BAA Comet Section, said it '. . . added to the mystery surrounding the bright apparition in August.'[14] Up till then all the observations had reference to an object with westward motion, but the Kershaw object suggested eastward motion. Of course the value of this report depended critically on the alignments used by Gaythorpe, and the accuracy with which the circumstances of the original observation were enacted. Gaythorpe qualified his result:

Any error there may be in the position of the observer can have no sensible effect on the altitude thus determined. Its effect on the azimuth, though much greater, is not, I think, likely to exceed $\frac{1}{2}$° at the most, with probably a greater tendency to decrease, than to increase, that coordinate. The observed position of the body relative to Capella tends to confirm the above value of the azimuth as approximately correct.[15]

Significantly, Mercury was about 8° from the place of the object. In view of the difficulties under which the observation was made, could it be that the young astronomers confused the planet for something else? More to the point, after three months were they able to recall the observing site in all its relationships?

The second observation was announced in Harvard College Observatory *Bulletin* No 759:

Bright Object near Sun. In a letter received at this Observatory, Dr H. C. Emmert of 3403 Warren Avenue West, Detroit, Michigan, states that he saw a bright object in the western sky on August 6 at 5h 50m p.m., Eastern Standard Time. "Its altitude was 14.5 to 15 degrees, its azimuth 85 degrees; the Sun's altitude was 15 degrees and azimuth 90 degrees. The object was fully as bright as Venus at her greatest brilliancy, and its light was perfectly steady." It is supposed that this object is the same as that seen near the Sun by five observers at the Lick Observatory on August 7, as reported in Bulletin 757.[16]

Harvard released the full text of the above letter, and Professor Campbell published it together with the other reports as a tail-piece to his report. It puts the observation in a different context:

> Reading an article by Dr. Russell in *Scientific American* for September 6, 1921, about the new object observed at Lick Observatory, I wish to give you my observation of August 6, 1921, 5:50 p.m. Eastern Standard Time. Having been deeply interested in astronomy for the last twenty-five years I casually glanced at western sky at that time and saw a very bright (silvery-white) object. Not being familiar with the position of *Venus* on that date I thought it was *Venus*, and yet it shone altogether too bright for *Venus*; after further observing I came to the conclusion it must be *Venus*. . . . The Sun at this time was obscured by buildings and the object was five to seven degrees east of Sun, and would say as many degrees south of ecliptic, the Moon being in sky at the time. At this hour in broad daylight this was a striking object. Did not scintillate. This must have been the same object seen at Lick Observatory three or four hours later, although they report it much nearer to Sun. This may be due to an enormous speed. A relative, Miss Crow, of Jackson, Michigan, asked me the name of the very bright star setting in the west about 7:30 p.m. the same date; this fixes the observation and date clearly in my mind.[17]
>
> (*Signed*) H. C. Emmert, M.D.
> Detroit, Michigan.

Campbell queried the date of the observation, but Emmert confirmed it was so. Yet this conflicts with his remark, 'This must have been the same object seen at Lick Observatory three or four hours later.' That he was confused over the date is clear. The manner in which he recalled the event adds to the suspicion. It is also strange that in noticing a conspicuous object in the heavens, despite his professed interest in the subject he singularly failed to verify his belief by reference to an almanac.

SUN-GRAZER OR NOVA?

Unlike the flare of a bright meteor which attracts even the most jaundiced observer, the fitful gleam in the sunset sky of August 7 1921 fled unheeded, except by those trained to recognise the unusual in such events. It is regretted that their reaction had no duplicate and that data are so scant. Other random observers tracked it briefly in a cloud-strewn sky towards and at sunset, in poor conditions. Doubtless they pondered its possible identity and character, but it appears they found it less extraordinary than did their professional counterparts. Consequently the record of its passage is rough, circumstantial, and contradictory, and based largely on memories revived by the publicity afforded the Mount Hamilton report. Complete evaluation, therefore, is not possible.

If the Lick and Plauen objects were identical, Crommelin suggested, '. . . and if we assume parabolic motion, the distance from the Earth did not exceed twice that of the Moon.'[18] But could this be? Of the five observations adduced, the two English reports of August 7, at 7 pm, and 8.30 pm, GMT, undoubtedly refer to the same object, and in all probability this corresponds with that seen from Mount Hamilton. The German report, on the other hand, though nearly simultaneous with them, at 7.35 pm, GMT, put the object almost 20° east of the Sun, whereas the English observers estimated its distance in the same direction at only 4° and 6° respectively. But, as we have already stated, Jupiter was near the Plauen position. And yet, as Professor Campbell noted: 'At the time and place stated the Sun would still be above the horizon and the object referred to must have been very bright to be visible.'[19] A fact that discounts the idea just put forward.

There is some possibility that the Emmert and Kershaw papers provide further evidence of the object, but the confusion over date in the first, and the alignment uncertainty in the second, do undermine their reliability.

Even so, the Kershaw evidence is interesting inasmuch as it gives a possible clue as to the character of the apparition. In the discussion that followed the reading of Gaythorpe's paper at the November 1921 meeting of the British Astronomical Association, Crommelin pointed out that if all the observations relate to real celestial bodies, that described by Gaythorpe indicated eastward motion, whilst the rest movement towards the west. Outwardly this suggests two bodies but the astronomical editor of *Nature*, in its columns of September 8 1921, commented that a comet with retrograde motion, moving near the plane of the ecliptic, and having small periheliom distance approaching the Sun from behind might remain in close proximity to it throughout the whole period of its apparition.[20] Such a comet if observed close to the horizon, as in the case of the Lick object, would assume the appearance of a bright star, since only its nucleus would be visible. Unfortunately, the estimates of distance from the Sun are all too rough to use for the deduction of motion. Of which Campbell remarked:

> If the object is a comet, its position could not be predicted because the orbit is totally unknown. It might be too near the Sun, either between Sun and Earth or on the far side of the Sun.[21]

Campbell and Russell both favoured the cometary hypothesis in the August 7 telegram.

E. E. Barnard unhesitatingly affirmed his opinion in a letter to Campbell, published by the latter:

> The case of this comet is not exceptional, as was shown by the Tewfik comet at the total eclipse of May 17, 1882, which was seen and photographed in Egypt, near the Sun [during total eclipse]. I, among others, hunted faithfully for that comet morning and night for a long time. Of course if the present comet was a sun-grazer it would move in a very narrow orbit, the whole of which might lie in daylight. It may be discovered yet in the southern hemisphere.[22]

Others took up the idea. Writing in the *English Mechanic*, September 2 1921, Colonel E. E. Markwick in congratulating Mr Fellows on his observation, also added, 'there is, to my mind, no reasonable doubt that it was a comet that was seen.'[23] Crommelin acknowledged the probability in his annual report of cometary activity for 1920–1, and ventured to designate the object 1921e.

Nature herself conspired to further this belief. Coincidental with the passage of the object, there were reports of unusual happenings in the nocturnal skies of Europe. A cablegram allegedly from Heidelberg, Germany, amplified and widely proclaimed in the popular press around August 10, stated that at Königstuhl the Earth had been observed to pass through the tail of a comet on the night of August 8. The *English Mechanic* August 19 1921, summarised the report thus:

> The Baden State Observatory at Königstuhl, near Heidelberg, reports that the earth passed through the tail of a comet during the night from Monday to Tuesday (8th–9th inst.). A number of luminous bands lay across a clear sky in the form of a wreath stretching from W.N.W. to E.S.E. The bands moved slowly in a N.N.E. direction, growing paler as the dawn came. The head of the comet passed southwards between the earth and the sun. It is suggested that the phenomenon has never been observed before, and that it was caused by the same body as that reported from the United States to have been visible in the vicinity of the sun at sunset on August 7.[24]

Although the report was incorrect, there was an element of truth in it. According to *Nature*, September 8, Dr Max Wolf, at Königstuhl, had seen on August 5 '. . . a long, very bright cloud west of the Pleiades, brightest near delta Arietis.' First noticed at 11h 15m GMT '. . . it faded rapidly, only a trace visible at 11h 36m.'[25] It was similar observations at 12h GMT, on the 8th, that formed the basis of the cablegram. This time the appearances were also observed at the Sonneberg Observatory, and no doubt this played an important part in what subsequently transpired. Be this as it may, Dr Wolf apparently did not send the message, nor did he intimate anything remotely interpretable as its content. Campbell suspected it came from the Wolff News Agency. Full details of the observations themselves can be found in the *Astronomische Nachrichten*.[26] Of related interest the observation by the English amateur Miss A. Grace Cook on August 4 1921 is worth quoting. As she wrote to the *English Mechanic*, Monday, August 8:

> I should very much like to hear from anyone who saw the luminous night sky on August 4. I went out to observe meteors at 9.30 GMT, and at once noticed a bright patch about 2 degrees

square, 10 degrees preceding beta Pegasi. I mapped its course, as I thought it might be the after-streak of a bright fireball I had missed. I found it was moving very slowly towards the North, therefore not in the prevailing wind, which was west and strong. Dark clouds were passing over from the West at the time. The little patch was visible two minutes. At 10.34 I altered my mind as to the appearance I had seen earlier, for I saw a long streak of very thin white cloud, not so bright as the Milky Way, but more like the Zodiacal Light, and the stars shone through it. The outer edge of the streak was touching alpha Pegasi at 10.34, and had reached alpha Andromedae by 11h., by that time the cloud appeared thinner and more difficult to see. I then saw other streaks stretching across the South and coming up gradually, lengthwise on, with dark gaps between. This continued till 13h., when I came in. I have often seen these luminous clouds or mist, but seldom so well defined.[27]

Noctilucent clouds? Or, as Crommelin and Wolf believed, auroral streamers? Or was there a connection between the bright object and the luminous sky phenomena? In short, did the gas and dust of a comet's tail graze the Earth in August 1921? Unlikely perhaps, yet it was a remarkable coincidence. And this raises the question posed by Barnard. If the object was indeed a comet, did it belong to that reckless family quaintly known as the Sun-grazers? If not, then what other explanation can be offered? There seems no doubt that a genuine celestial apparition was observed. Campbell disposed of the nova hypothesis:

If the object was a new star, it should have been observable Monday evening unless the brightness was greatly reduced. In the latter case there will be little chance of observing the object in the bright sky surrounding the Sun, for several weeks to come. The position of the object, R.A. 9h 22m., Dec. $+15\frac{1}{2}°$, about 40° from the galactic plane militates against the nova hypothesis.[28]

If not a nova, or a comet, what? A member of the minor planet group that occasionally pass very near the earth, perhaps?

ADDENDA: Since the foregoing was written Dr Joseph Ashbrook, editor of *Sky and Telescope*, has written up the story. He states how Dr Brian G. Marsden, Smithsonian Astrophysical Observatory, has checked the possibility advanced by Barnard that the object was a Sun-grazer. Marsden's calculations indicate that if the object were a comet, then it could not belong to this famous group. The computations were based on the assumption that because all the known members follow closely similar orbits, then so should the object. This is reasonable, but is it not perhaps possible for other unknown members to follow orbits with different characteristics?[29]

5 William Herschel and the Ring of Uranus

> *I will make such telescopes, and see such things!*
>
> William Herschel (1782)

UNLIKE his contemporaries, who observed the heavenly bodies usually to confirm their place on the celestial sphere, William Herschel embarked upon his sky reviews, as he called them, in the manner and spirit of an explorer. He searched incessantly for whatever could be found in the heavens, resolved 'to take nothing upon trust,' as he once wrote, 'but to see with my own eyes all that other men had seen before.'[1]

Four distinct times in his life did Herschel thus review the heavens in long nightly vigils. Each lasted several years, and encompassed the entire northern sky accessible to his telescopes. His purpose: to scan all parts without preference or preconceived ideas, and to study everything that came within the light grasp of his telescopes. Eighteenth-century astronomy had been chiefly concerned with the mechanics and stability of the Solar System. Herschel changed all that and re-orientated attention to the remote realm of the nebulae.

MARCH 13 1781

Mature considerations aside however, in the popular mind Herschel is best remembered for the notable by-product of his second sky review. Accomplished with a telescope less than 7 feet in focal length, and $4\frac{1}{2}$ inches aperture, the first review had convinced him of the need to implement his programme with more powerful optical means. Accordingly he undertook the construction of a larger telescope. Completed in 1778, of the Newtonian type, it had an aperture of slightly more than 6 inches, a focal length of about 85 inches and was armed with a power of 227. With it Herschel inaugurated his second review in August 1779. This had a limiting magnitude of 8, and included a search for double stars suitable for the purpose of determining the stellar parallax by the method suggested by Galileo.

Late on the night of Tuesday, March 13 1781, whilst engaged in this pursuit, Herschel made the discovery destined to revolutionise his circumstances. Between 10 and 11 pm, he was sweeping through the western extremity of Gemini, when amongst some faint stars near H Geminorum (now identified as 1 Geminorum) he perceived an object sensibly larger than its neighbours, which gleamed with a steady bluish light. Comparison with H Geminorum and another faint star close by, possibly the fifth magnitude star 132 Tauri, confirmed its exceptional size. Substituting the low for higher magnification Herschel was astonished to find the object increased in size, its image assuming a hazy, indistinct aspect. Instantly he realised this was no ordinary star, and in his first manuscript notation he stated he had seen, 'a curious either nebulous star or perhaps a comet.' Its motion discounted the former idea. Struck by its obvious peculiarity, and anxious to pursue it before it was lost in the solar rays Herschel continued his watch until April 19. He then prepared an account of his observations and mistakenly announced to the Royal Society he had discovered a new comet.

On this assumption various mathematicians tried to compute its orbital elements and compose an ephemeris. To their frustration, failure confronted every attempt. For a few days observed and theoretical places would coincide, then inexplicably diverge. Many tedious hours were thus spent in trying to reconcile theory with observation. Only after much laborious effort had been spent did the true character of the object begin to emerge. The breakthrough came when de Saron announced to the French

Academy of Sciences on May 8 1781 that in reality it lay much further from the Sun than everyone thought, and gave as its perihelion distance a value not less than twelve times the radius of the Earth's orbit. Anders Johann Lexell took the initiative. Utilising two extreme observations, one by Herschel of March 17 and the other by Maskelyne taken May 11 1781, Lexell calculated the elements on the basis of circular motion. 'From that time,' Lalande wrote, 'it appeared to me that the body ought to be called the new planet.'

THE RING ENCOUNTERED

Seven years later Uranus, or the Georgian as it was then known, subjected Herschel to an unusual embarrassment, the details of which may not be familiar. It is the imprint of a period when he struggled to distinguish something of which he could not be certain. Eventually he decided it had been a deception.

Everything has its beginnings, and this incident is no exception. It all started when Herschel with his small telescopes sought to inquire of the planet's physical condition as evidenced by its surface detail and phenomena, and also to determine if it had a satellite.

Uranus shines with a blue-green light, more blue than green. If its position is known, and atmospheric conditions allow, it may be seen with the naked eye. Small telescopes reveal a neat, well defined disk. Its immense distance, however, precludes all observation of its more intimate appearances. None the less Herschel persevered, and from observations with his 7, 10 and 20-foot reflectors, considered its shape to be spherical, and its limb sharply etched. Though he repeatedly examined the planet and its environs, the result was disappointing, the observations of one night contradicting those of the next. Indeed in April 1783 his earlier impression could not be verified, and he surmised Uranus showed a degree of polar flattening suggesting therefore affinity with Jupiter and Saturn. He ascribed his failure to insufficient light grasp, and for a time other objects diverted his interest.

Early in January 1787 he applied the front-view technique to his observation of the nebulae, and noted with surprise how much brighter they appeared compared to his former views with the Newtonian, and moreover how more distinct. Aided by this different optical arrangement he decided:

> It would not have been pardonable to neglect such an advantage, when there was a particular object in view, where an accession of light was of the utmost consequence; and I wondered why it had not struck me sooner.[2]

Consequently, on January 11 1787 in the course of his general review he selected a particular sweep 'which led to the Georgian planet.'[3] As it culminated he perceived in its immediate vicinity several faint stars, the configuration of which he depicted with great care. The next night as the planet ascended once more to the meridian, it again attracted his attention. Instantly he noted two of the stars were missing. Suspecting them to be satellites, but wary of optical defect, or possibly a subtle condition of the atmosphere which momentarily hid them from sight, he judged that nothing short of a series of observations would resolve the enigma. To this end he charted all the stars around the planet as he found them on January 14, 17, 18 and 24, and also on February 4 and 5 1787. He sensed the reality of at least one satellite, but deferred any announcement until he had actually seen it in motion.

The night of February 7 gave him the opportunity he needed, and at 6 pm, he directed his 20-foot telescope towards Uranus. Through long hours he followed it steadily until 3 am the next morning, 'at which time, on account of the situation of my house, which intercepts a view of part of the ecliptic, I was obliged to give over the chase.'[4] In the nine hours he maintained his vigil, one tiny brilliant faithfully adhered to the planet, but more significantly it had shifted over a substantial arc, thus betraying its real character.[5]

Whilst engaged in this pursuit, Herschel did not neglect to track a second tiny star which from its elusive behaviour he also suspected of being a satellite: first seen on January 11, missed on the 12th; recorded on the 14th, marked absent on the 17th; to reappear on the 18th, only to have vanished by the 24th. On February 7 Herschel caught a brief glimpse of it, but the glare of the planet partially concealed it from sight. His preoccupation with the other satellite, however, prevented his obtaining satisfactory confirmation of its existence. Moreover, towards morning, when a change in its position would have become evident, moonlight encroached upon the field and effaced it from view.

Undaunted, Herschel returned to the pursuit on February 9. His first satellite lay exactly where he supposed it would, the second some further distance from where he had seen it last. It had travelled in the same direction as the first satellite, yet at a faster rate indicating it lay nearer the planet. Though the last to be discovered, Herschel enumerated this the first satellite, and the other the second, now known respectively as Titania and Oberon.[6]

Now it happened that during the course of these observations Herschel noted in his journal under date of February 4 1787, '20 feet reflector, power 300. Well defined; no appearance of any ring, much daylight.'[7] Whatever is to be made of this entry, whether he had seen traces of annulus about the planet, or merely sought to distinguish features analogous to those seen in the other known planets, we cannot affirm. However it is regarded, this stands as his first reference to the alleged ring of Uranus.

On March 4 1787 he found the planet somewhat flattened at the poles, but when most clearly seen, attended by four projecting points of light, almost at right angles to each other, which in his view resembled, 'Perhaps a double ring; that is, two rings, at rectangles to each other.' (Fig 2.)[8] A cursory glance at Jupiter the following night disclosed suitable seeing conditions, and he again turned to his planet to find, '. . . it was again seen affected with projecting points. Two opposite ones, that were large and blunt, from preceding to following; and two others, that were small and less blunt, from north to south.'[9] By the 7th, the idea of a ring was implanted for he reported, 'Position of the great ring R, from 70° S.P. to 70° N.F. Small ring r, from 20° N.P. to 20° N.F. 600 shewed R and r. 800 R and r. 1200 R and r.'[10] But

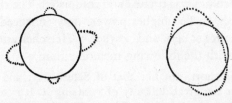

Fig 2 (left) Double ring of Uranus as seen by Herschel,
1787 March 4; *(right)* single ring as depicted by Herschel,
1789 March 16.

on the 8th, Herschel entered into his record, 'R and r are probably deceptions.'[11] He verified this belief on November 9, 'The suspicion of a ring returns often when I adjust the focus by one of the satellites, but yet I think it has no foundation.'[12]

Yet almost a year and half later, on February 22 1789 to be precise, he wrote, 'A ring was suspected.' He continued. March 16 at 7h 37m, 'I have turned my speculum 90° round. A certain appearance, owing to a defect which it has contracted by exposure to the air since it was made, is gone with it; but the suspected ring remains in the place where I saw it last.' (Fig 2.)[13] From his sketches illustrating the 'certain appearance', Uranus had

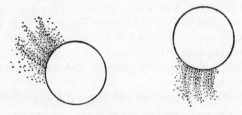

Fig 3 Defect in the telescopic appearance of Uranus described by Herschel, 1789 March 16. (See text.)

seemingly developed a flare-like pattern on one side, and this sounds suspiciously like the image defect called *coma* (Fig 3). Often this is caused not by faults in the optical system, but by incorrect squaring on or alignment of the optical elements. The fault lies in images on either side of the optical axis of the objective. And this explanation is confirmed by the fact that the fault disappeared when the speculum was rotated and readjusted. The ring however did not vanish. In fact higher powers only enhanced its appearance at 7h 50m ($\times$ 471 and $\times$ 589).[14] Herschel annotated this observation with the following memorandum:

> The ring is short, not like that of Saturn . . . and this may account for the great difficulty of verifying it. It is remarkable that the two ansae seem of a colour a little inclined to red. The blur occasioned by the fault of the speculum is, to-night, as represented . . . the ring is likewise as it was the same evening.[15]

It is plain that at this period Herschel's optical arrangement was plagued by fault and imperfection, and this is emphasised by his observation on March 20 at 7h 53m, 'When the satellites are best in focus, the suspicion of a ring is the strongest.' December 15, 'The planet is not round, and I have not much doubt but that it has a ring.'[16]The change from the plural to the singular is interesting in these later reports of the annulus, and underlined Herschel's growing suspicions of its nature.

Nearly three years later, February 26 1792 at 6h 34m, he was still undecided, though:

> My telescope is extremely distinct; and, when I adjust it upon a very minute double star, which is not far from the planet, I see a very faint ray, like a ring crossing the planet, over the centre. This appearance is of an equal length on both sides, so that I strongly suspect it to be a ring. There is, however, a possibility of its being an imperfection in the speculum, owing to some slight scratch: I shall take its position, and afterwards turn the speculum on its axis.
>
> 8h 39m. Position of the supposed ring 55°.6 from N.P. to S.F.

Adjustment of the speculum did little but render the matter all the more perplexing:

> 9h 56m. I have turned the speculum one quadrant round; but the appearance of the very faint ray continues where it was before, so that the defect is not in the speculum, nor is it in the eye-glass. But still it is now also pretty evident that it arises from some external cause; for it is now in the same situation, with regard to the tube, in which it was $3\frac{1}{2}$ hours ago; whereas, the parallel is differently situated, and the ring, of course, ought to be so too.[17]

He examined the planet further on March 5 with a newly polished mirror, but though he applied successively higher power (from × 240 to × 2400), all of which were borne tolerably well, the ring did not appear. Moonlight pervaded the telescopic field however, so that the ring would have been very difficult to resolve in any case.

'Some apparatus about the planet,' was noted on December 4 1793, but on this occasion the image appeared strangely indistinct and not so sharp as it ought to have been considering 'the extraordinary distinctness of my present 7-feet telescope,' as he commented. The appearance of the ring was nothing more than a suspicion. Observations in February 1794 and April of the following year omit any mention of this feature.

And that terminates Herschel's record of the ring. James South (1785–1867) in describing the performance of his 20-foot achromatic May 14 1830 remarked:

> *Georgium Sidus.* At half-past two, placed the 20-feet achromatic on the Georgium Sidus; saw it with 346, a beautiful planetary disk; not the slightest suspicion of any ring, either perpendicular or horizontal; but the planet three hours east of the meridian, and very low. Morning beautiful, but moonlight very strong, and the moon within three degrees of the planet.[18]

In a paper written on the eve of his departure for the Cape of Good Hope in November 1833, and communicated to the Royal Astronomical Society, Sir John Herschel commented:

> Neither has he [Sir John] ever seen any appearance about the planet which gives ground for the least suspicion of a ring.[19]

Whilst A. A. Common in a letter to W. F. Denning quoted by the latter in his *Telescopic Work* (1891), stated:

> With only moderate powers, Uranus does not show a perfectly sharp disk. No markings are visible on it, and nothing like a ring has been seen round it.[20]

Common observed in the spring of 1889, possibly with his large reflector.

CONSIDERATIONS AND CONCLUSIONS

When the suspicion first arose in 1787, Herschel anticipated that the ring, 'might be in such a situation as to render it almost invisible;' that is, its plane was obliquely inclined to the earth. Consequently Herschel had decided, 'observations should not be given up, till a sufficient time had elapsed to obtain a better view of such a supposed ring, by a removal of the planet from its node.' Thus no matter what the inclination of the ring had been at its discovery, provided it had kept the same relationship with the planet, after ten years (Herschel dated his paper September 1 1797) not only should its aspect have varied but its disposition in regard to the Earth should have yearly become more favourable. But the conjecture had not been fulfilled. Expressing great confidence in his observation of March 5 1792, when he used a newly polished speculum and found Uranus affected solely by a slight compression at its poles, this aspect being supported by later observations, Herschel ventured to affirm that Uranus 'has no ring in the least resembling that, or rather those, of Saturn.'[21] Obviously he had finally realised that deception had interceded. Additional evidence to this effect was supplied by his observation of February 26 1792. On that occasion the ring in $3\frac{1}{2}$ hours had not enjoined in the motion of the planet itself, but throughout that interval had remained stationary with regard to the telescope tube. A fact decisively against the existence of the ring.

Atmospheric disturbance plays havoc with telescopic definition, and the larger the telescope the more pronounced are its effects. Next to cloudiness it is the most serious hindrance to astronomical observation, and often results in illusory appearances. Yet Herschel was too astute to be so deceived, and an explanation framed in these terms must be rejected. This leaves only one alternative; image defect produced by instrumental imperfection. 'With regard to the phaenomena which gave rise to the suspicion of one or more rings,' Herschel observed in his discussion, 'it must be noticed, that few specula or object-glasses are so very perfect as not to be affected with some rays or inequalities, when

high powers are used, and the object to be viewed is very minute.'
His experience of March 16 1789, and earlier, serves as a good
example. For the flare-like pattern observed to one side of the
image of Uranus, as described and sketched by Herschel, is not
inconsistent with the idea that the optical components of his
telescope were slightly out of alignment at that time. But this he
thought would not explain the formation of the ring, its cause
'must be looked for elsewhere.'

Although the speculum metal mirrors of those days gave a
reasonable performance, they only reflected about 60 per cent of
the incident light, which compares unfavourably with modern
mirrors. Worse still the intrusion of the diagonal and the eyepiece
instituted a further reduction. Inevitably image quality was poor
by today's standards.

Now in 1728 Le Maire had described to the French Academy
of Sciences an idea that promised a practical solution to the
problem. In short he advanced an entirely new concept in telescope construction. He proposed the main mirror should be tilted
at a small angle to the optical axis, so that it reflected the con-
vergent cone of rays direct to the eyepiece mounted on the upper
edge of the tube. Naturally the observer would stand with his back
to the object under study. Often credited to Herschel, this arrang-
ment is generally referred to as the Front view.

Herschel had been attracted to this notion in 1776, and in 1783
and 1784 had conducted a number of experimental trials in its
construction, but had not immediately followed them up. When
he did in January 1787 the advantages were obvious. Not only did
it facilitate his observations and enable him to make them in comfort, relatively speaking, but the light gained by the suppression
of the diagonal was especially noticeable in his views of the
nebulae, as we have noted. Generally the experiment had been a
success, but in its wake came the ghost ring. Though he suspected
its character, he could not define its precise cause. But already, as
he wrote in 1797:

> It has often happened, that the situation of the eye-glass, being
> on one side of the tube, which brings the observer close to the

mouth of it, has occasioned a visible defect in the view of a very minute object, when proper care has not been taken to keep out of the way; especially when the wind is in such a quarter as to come from the observer across the telescope. The direction of a current of air alone may also affect vision.[22]

It would appear that Herschel had made acquaintance with the characteristic defect of the Front view, namely astigmatism. This is common to mirrors and lenses, and to the eye, and is caused by a difference of curvature between mutually perpendicular planes, which means that whereas a ray of light in one plane may be in focus, a similar ray in the other plane will not, consequently the size and shape of an image will vary for different focal points inducing a visible defect in the image. An astigmatic image is linear, an appearance that coincides with Herschel's description and delineation of the Uranian ring.

The Front view has nowadays fallen into disuse because of the rapid introduction of this type of defect at large off-axis distances. Telescopes of low focal ratio if designed in this fashion are particularly subject to it, hence the basic reason for the demise of the method since modern work relies on short focus instruments in the main. On the other hand, if the focal ratio is high the distortion is proportionately reduced. To a degree, Herschel inadvertently minimised its effect by using mirrors of very long focal length, which enabled him to tilt them off-axis without appreciably impairing image quality. Moreover it seems entirely possible that the figure of the $18\frac{1}{2}$-inch mirror was suitable for use in this way, for Herschel inclined to use high magnification, up to 7,000. The probability is that this was his only mirror that would efficiently perform off-axis. Indeed it is significant that it was this instrument that John Herschel took to the Cape of Good Hope in 1833 for his great survey of the southern heavens, not the great 48-inch reflector, the performance of which was so poor that its use was denied to any astronomer.

It is interesting to compare this episode with the details that make up the affair of the Neptunian ring, which we shall now consider. Astigmatism may have been its cause, but the 'discovery'

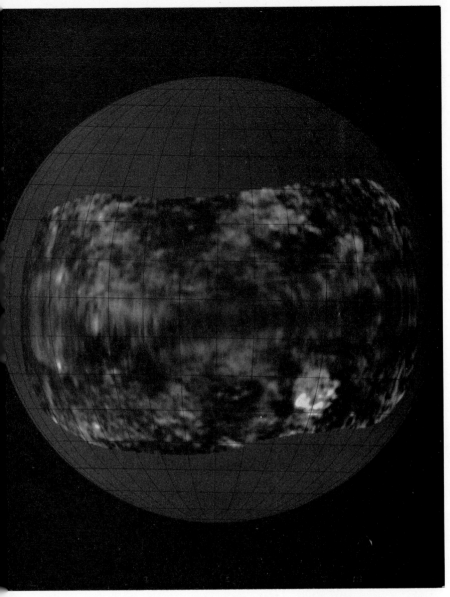

Page 117 Radar Map of Venus by R. M. Goldstein and H. Rumsey Jr; an enlarged version of (c) Plate p 68. About 30 million square miles of the surface is depicted, covering the Equatorial region. Most of the north-south ambiguity has been removed. North is at the top.

(43) Starfield 19th Jan 1847

My dear sir

In reply to your note
of the 16th I deeply regret to say
that such has been the unva=
rying cloudiness of the weather
in this district that not a sin=
gle opportunity has occurred
this year (nor indeed for some
time previous to its commence
ment) of seeing the new planet.
The amount of evidence of
my having seen a supposed
ring is very correctly recorded
in the minutes of the december

Lassell

Page 118 The ring of Neptune as depicted by William Lassell from observations with his 24-inch reflector during the autumn and early winter 1846. Fragment is from his letter to Professor Challis, Cambridge, of date 1847 January 19. Original in the archives of the Cambridge Observatories.

had wider repercussions. There are, however, some points of similarity in the two episodes, and most probably they share a common origin.

6 Neptune 1846-7

A ring was suspected.
William Herschel (1789)

IN our review of the Uranian ring, we sketched in the case history of an optical effect peculiar to one observer, and attributable, in all likelihood, to the type of mirror system employed. With the alleged ring of Neptune, on the other hand, the circumstances of its accession were totally different, and exclude comparison with Herschel and his observations. For in this episode not one, but several reputable observers unanimously agreed that when most steadily seen Neptune seemed to be affected by a faint and luminous feature, of symmetrical form, which assumed in moments of atmospheric calm, the shape of a ring obliquely inclined to the line of sight. It was thus seen equally well on numerous occasions, under various conditions, with both forms of telescope objective. The details of its history are direct and uncomplicated, and in the first instance run parallel to those of its Uranian counterpart. Yet of the phantom to which they relate, no satisfactory explanation can be offered. That illusion invoked its exhibition is certain, as modern observation has failed to attest its existence, but other factors complicate the diagnosis; factors of some perplexity, not easily comprehended other than by vague allusions or generalisations, and which, we suspect, have their origin in a basic human weakness—presuggestion.

The facts here collected have been taken from sources not generally known, or accessible, and in several instances from unpublished correspondence.

THE DISCOVERY OF NEPTUNE

The optical discovery of Neptune took place on the night of September 23 1846. It marked the close of a long period of speculation and mathematical inquiry in direct interpretation of the law of gravitation, and is rightly acclaimed to be one of the greatest triumphs of theoretical astronomy. By way of preface the main facts of this classic chapter are here recalled.

Working on improved tables of Jupiter, Saturn and Uranus in 1820, Alexis Bouvard (1767–1843), of Paris, found that although Jupiter and Saturn moved in obedience to the gravitational principle, Uranus did not. Even after taking into account the perturbations of all the known planets, he failed to reconcile the pre-1781 observations with those made in the early part of the nineteenth century. Rejecting the old observations as unreliable, he based his tables on the modern positions, and left to posterity the awkward decision as to whether the discrepancy were due to inaccurate observation on the part of the older observers, or to the action of some extraneous agency.

Only a few years elapsed before Uranus again began to deviate from its calculated orbit. By 1830 the anomaly amounted to some 20 seconds of arc. Ten years later it had risen to 90 seconds of arc, and continued to increase until by 1845 it had reached the so-called 'intolerable quantity' of nearly 2 minutes of arc. To the uninitiated such minute quantities may appear insubstantial, almost to the point of insignificance, and if translated into the actuality of unaided visual observation they may well feel justified in this belief. If we suppose two bodies, one in the place of the real planet, the other in that of the fictitious or theoretical, then it would have required an eye of exceptional keenness to distinguish them apart, at least during the early part of this period. Magnified by the telescope the outwardly small becomes measurably large, and is not for a moment to be neglected. In exact science such unexplained though unquestionably true residuals are more often than not the very essence of discovery. In this particular instance the residuals supplied the data from which the position of another majestic world was disclosed, still farther afield than Uranus.

Naturally the probable cause of these anomalies became a source of speculation and discussion, and many suspected the true nature of the disturbance long before it was finally recognised in the telescope. Even so, no one felt particularly inclined to take up the challenge until around 1844.

In February of that year, after a trial investigation, John Couch Adams (1819–92), a young Cambridge mathematician, applied through James Challis (1803–82), Plumian professor of astronomy at Cambridge, and Director of the University Observatory, to George Biddell Airy (1801–92), the Astronomer Royal, for information on the irregularities in the motion of Uranus. Adams intended, as he wrote in his diary of the preceding year, to produce a definitive solution by computing the elements of the hypothetical planet, and to estimate its mass in terms of the Sun, on the assumption that it was twice as distant from the Sun as Uranus; a premise he based on the rough empirical relationship between the mean distances of the planets, popularly known as *Bode's Law*, but more accurately the rule of Titius. He completed this task by October 1845, and deposited with Airy data so near the truth that if a search had then been made about the place indicated, the planet would hardly have escaped detection.

But incredulity crept in, and though Airy did acknowledge the papers he deferred a search on the pretext of a slight technicality which he raised with Adams. To this Adams gave no reply. At this point the story is marred by indifference and tardiness, matters too prolix to form part of this narrative. In any case it would go beyond our allotted span to relate the unfortunate happenings and non-happenings that delayed the publication of these results. Suffice it to say nothing was immediately published on the subject and no attempt made to secure Adams' right to priority.

That same year the problem engaged the interest of a rising young French mathematician, Urbain Jean Joseph Le Verrier (1811–77). Unaware of the Cambridge investigation, and acting upon a suggestion of his friend and colleague François Arago, he began a critical examination of Bouvard's tables to ascertain if the errors were indigenous to their theory, and recalculated the perturbations of Uranus produced upon it by Jupiter and Saturn. Apart from a number of minor errors, Le Verrier found nothing to satisfy the observed anomalies in the motion of Uranus.

After making full allowance for the disturbing effects of Jupiter and Saturn, Le Verrier next tried to assign a new orbit but without success, the best orbit showed deviations too great to be attributed to errors of observation. He concluded, therefore, that the discrepancies were due solely to the attraction of an unknown body, situated beyond Uranus probably at twice its distance from the Sun; this being consistent with the well known 'law'.

Le Verrier communicated three memoirs on the question to the French Academy of Sciences. In his first paper, submitted November 10 1845, he proved that the anomalous movement of Uranus could not be accounted by reference to any known disturbing agent. With the second, presented June 1 1846, he established that the anomalies indicated the existence of a body beyond Uranus, of which he estimated the probable position. In his third and final memoir, given on August 31 1846, he stated the elements of the hypothetical planet (data which define the size and shape of an orbit, its orientation in space and the position of the body concerned for a specific epoch). In his paper Le Verrier stressed the one fact that expedited the discovery. His calculations of its mass confirmed that the planet would show a discernible disk, and that it would be sufficiently bright to be conspicuous in ordinary telescopes.

Captivated by the prospect, the astronomical fraternity awaited with suppressed excitement the discovery that would inevitably follow this announcement. And here and there observatories began to make plans to search for the new planet.

Struck by the coincidence of these results with those of the ill-fated Adams, Airy discussed the matter with Challis and urged him to undertake a search with the large Cambridge refractor. Unfortunately, Challis adopted a method which although it made the discovery certain, was none the less extremely laborious. Instead of endeavouring to recognise the planet by the disk-method suggested by Le Verrier, he elected to find it by its motion against the stellar background. Photographic search techniques would nowadays quickly resolve such an issue, but in 1846 the only available way required the observer to chart and repeatedly measure all the stars in the suspect neighbourhood, so as to find which of them had changed its position. Since observations accumulated in large numbers, their reduction consumed inordinate amounts of time. Consequently the work of comparison proceeded but slowly. We need briefly remark, therefore, that the object of the search could be in the grasp of an observer some considerable time before he actually appreciated the fact. So it happened with Challis. As he carried out his dignified search, the planet twice passed through his telescopic field, on August 4 and 12 1846, only to be recorded as a star. If the observations had been immediately reduced, history would record the discovery of Neptune one month sooner than it does. As it was, other engagements delayed the numerical work; a circumstance that denied Challis and Cambridge the glory that immortalised others.

Meanwhile on September 18 1846, while Challis was still working away at his observations, still unconscious that Neptune lay secure in the pencilled notations of his records, Le Verrier himself wrote to Johann Gottfried Galle (1812-1910), then assistant at the Berlin Observatory, stating the spot in the heavens where the planet would probably be found, and inviting Galle to participate in the search; how and why Le Verrier came to select the German astronomer is fully described by Morton Grosser in his excellent book *The Discovery of Neptune* (Harvard, 1962), page 115.

Galle received this letter with enthusiasm on September 23. Johann Franz Encke (1791-1865), Director of the Observatory, did not share the same attitude, and at first refused Galle's proposal that they should conduct a search that very night. Galle persisted, and eventually obtained grudging approval from his superior. Heinrich Louis d'Arrest (1822-75), a student-astronomer at the observatory, chanced to interrupt this conversation. Sensing a variation in the normal routine of observatory life, he asked if he also could take part in the hunt for the hypothetical planet. Having secured permission to use the 9-inch Fraunhofer refractor that night, Galle felt disinclined to subdue the spirited response of the young man.

In the course of the day Galle computed the geocentric coordinates of the planet from the data supplied by Le Verrier, and in the evening both he and d'Arrest directed the telescope to the appointed spot. As Galle searched for the tiny disk, d'Arrest waited patiently, quiet and expectant. After a fruitless search, d'Arrest suggested the use of a star chart. But Galle was lukewarm about the idea; his most recent experience had taught him to mistrust their accuracy. Still the suggestion had merit and it might save time; d'Arrest rummaged around in the observatory map cupboard and produced a new chart of the very region they were sweeping. Armed with this Galle returned to the refractor, whilst d'Arrest sat himself down at a desk with the map spread out before him. So the process of identification began. Galle calling out the stars, d'Arrest carefully checking them against the stars on the map, until at last an eighth magnitude object came into view within seven minutes or so of the predicted position in right ascension. Excitedly, d'Arrest declared, 'That star is not on the map'. But was it the rumoured eighth planet? Soon after midnight, the news brought Encke hurrying round to the dome, and the three astronomers followed the object until it set about 2.30 am. Some doubt existed as to whether they had found the planet. Its disk had proved difficult to resolve, and its motion infinitesimal. The following night, September 24 1846, they again found the object. This time there could be no doubt, it had moved its place among the stars—indeed it was the eighth planet. By the early autumn it had been observed by many astronomers in America and Europe, and in particular by the well known English amateur William Lassell (1799–1880).[1]

WILLIAM LASSELL

Lassell, who commenced business as a brewer in Liverpool in 1825, 'without, however, much taste or inclination for trade,' spent the whole of his limited leisure time in the pursuit of his favourite interest, astronomy and the mechanics of telescope construction. Unable to purchase expensive instruments, yet desirous of obtaining a complete knowledge of celestial observation, he began to make his own telescopes in 1820, in his twenty-first year.

His success with a Newtonian of 7 inches aperture, and a Gregorian of the same size, attempted simultaneously, encouraged him to manufacture another Newtonian of 9 inches aperture which he fastened to an equatorial movement, on a plan of his invention.[2] In 1840 Lassell installed this telescope in an observatory at Starfield, his residence near Liverpool. He made several mirrors for this instrument, all of excellent quality. Of the many visitors to Starfield, one of the most frequent was the Reverend William Rutter Dawes (1799-1868), renowned for his visual acuity and skill as an observer of double stars. His observatory note-books, which passed into the possession of Sir William Huggins, abound with details of the many delicate tests for figure which Lassell applied to these mirrors to satisfy himself of their continuing high performance and quality. This telescope, incidentally, revealed to Lassell the sixth star in the trapezium of the Great Nebula in Orion (M.42), without his prior knowledge of its existence.

But Lassell was imbued with the bold spirit of the elder Herschel, and by 1844 had decided upon the construction of a 24-inch reflecting telescope, 242 inches in focal length, to be mounted on the same principle as the 9-inch Newtonian. He spared no effort to make this as perfect as possible, both in its optical and mechanical arrangements. As a first step he visited the Earl of Rosse at Birr Castle, ostensibly to satisfy himself on the performance of the great 36-inch reflector, but more importantly to examine the apparatus by which its mirrors were polished. Impressed by what he saw, Lassell resolved, upon his return to Starfield, to model his polishing machine along the same lines. Several months work with this device convinced him that its linear motion was less satisfactory than he at first supposed. Consequently he contrived a machine that simulated the movement of the hand, 'by which he had been accustomed to produce perfect surfaces on smaller specula.' Lassell introduced several new ideas and techniques into all aspects of his project, innovations too numerous to mention here. It is necessary, though, to remark that he had the good fortune to enlist the help of James Nasmyth

(1808–90), a Scottish engineer and inventor of the steam hammer, and also a fellow astronomer.[3] In this connection we can do no better than quote Sir John Herschel:

> . . . that in Mr. Nasmyth he was fortunate to find a mechanist capable of executing in the highest perfection all his conceptions, and prepared by his own love of astronomy and practical acquaintance with astronomical observations, and with the construction of specula, to give them their full effect.[4]

Installation of the 24-inch, then the largest reflecting telescope in England, began in 1845 and was completed in the following year. With it Lassell discovered Triton, the faint inner satellite of Neptune, in 1846; Hyperion, the eighth satellite of Saturn in 1848, simultaneously with William Cranch Bond (1789–1859), first Director of the Harvard College Astronomical Observatory; and in 1851, after a long and careful search, two faint satellites of Uranus, Umbriel and Ariel, both interior to those found by William Herschel in 1787. For the construction of this remarkable telescope and the discoveries made with it, the Royal Astronomical Society awarded Lassell its Gold Medal in 1849. In his presentation address, the President of the Society, Sir John Herschel, eulogised in this manner:

> The simple facts are, that Mr. Lassell cast his own mirror, polished it by machinery of his own contrivance, mounted it equatoreally in his own fashion, and placed it in an observatory of his own engineering. A private man, of no large means, in a bad climate and with little leisure, he has anticipated, or rivalled, by the work of his own hands, the contrivance of his own brain, and the outlay of his own pocket, the magnificent refractors with which the Emperor of Russia and the citizens of Boston have endowed the observatories of Pulkowa and the Western Cambridge.[5]

But there is another side to this record, about which Sir John said nothing. A set of observations, small in number, most of which are now lost, in part confirmed by others, and which in their import deposed Saturn from its unique place in the planetary system.

'I SUSPECTED THE EXISTENCE OF A RING'

In the week following the discovery of Neptune, the Berlin astronomers despatched a heavy correspondence to all the major astronomical institutes in Europe and America giving details of the observations and the planet. Airy learnt of it on September 29, just after he and his family arrived in Gotha for a three-day holiday. News reached London the day after, and a letter from John Russell Hind announcing the event appeared in *The Times* of October 1 1846.[6]

On Wednesday, October 14, that same newspaper carried the following letter:

Sir, On the 3d instant, whilst viewing this object (Neptune) with my large equatorial, during bright moonlight, and through a muddy and tremulous sky, I suspected the existence of a ring round the planet; and on surveying it again for some time on Saturday evening last, in the absence of the moon, and under better, though still not very favourable atmospherical circumstances, my suspicion was so strongly confirmed of the reality of the ring, as well as of the existence of an accompanying satellite, and I am induced to request you as early as possible to put the observations before the public.

The telescope used is an equatorially mounted Newtonian reflector, of 20 feet focus, and 24 inches aperture, and the powers used were various, from 316 to 567. At about $8\frac{3}{4}$ hours, mean time, I observed the planet to have apparently a very obliquely situated ring, the major axis being seven or eight times the length of the minor, and having a direction nearly at right angles to a parallel of declination. At the distance of about three diameters of the disk of the planet, northwards, and not far from the plane of the ring, but a little following it, was situate a minute star, having every appearance of a satellite. I observed the planet again about two hours later, and noticed the same appearances, but the altitude had then declined so much that they were not so obvious. My impression certainly was that the supposed satellite had somewhat approached, but I cannot positively assert it. With respect to the existence of the ring, I am not able absolutely to declare it, but I received so many impressions of it, always in the same form and direction, and with all the different magnifying powers, that I feel a very strong persuasion that nothing but a finer state of atmosphere is necessary to enable me to verify the discovery. Of the existence of the star, having every aspect of a satellite, there is not the shadow of a doubt.

Afterwards I turned the telescope to the Georgium Sidus, and remarked that the brightest two of his satellites were both obviously brighter than this small star accompanying Le Verrier's planet.[7]

I remain, Sir, your obedient servant,

WM. LASSELL.

Starfield, Liverpool, October 12.

Lassell supplied further details of his observation of Saturday, October 10 to the *Monthly Notices*:

The same impression of a ring in the same direction. A minute star just steadily visible, with full aperture of 24 inches, powers 316 to 567, distance 2½ to 3 diameters, a little to the right, and *apparently below* the ring continued. Speculum B and Merz's prism.[8]

On November 10, Lassell found 'The planet very like Saturn, as seen with a small telescope and low power, but much fainter. Same speculum and prism as before.' The next night, 'The planet still retains its appearance. A faint point of light considerably distant, in the direction of the ring and below it. Speculum A and plane reflector.' Visitors to Starfield independently confirmed this configuration in every aspect, although Lassell neglected to say who these persons were.

Unfortunately, conditions were not conducive to good observation. For some time there had been an unusual prevalence of cloud, and a general unsteadiness of seeing. Moreover, the unfavourable position of Neptune south of the celestial equator, in the border region of Aquarius and Capricornus, frustrated every attempt to determine the character of this ghostly feature.[9] What with the haze and disturbed air near the horizon, and the effects of differential refraction, Lassell experienced much annoyance in his observations of Neptune, its satellite and ring, as he complained to Challis in a letter dated Starfield, January 19 1847 (see Plate, page 118):

> . . . I deeply regret to say that such has been the unvarying cloudiness of the weather in this district that not a single opportunity has occurred this year (nor indeed for some time previous to its commencement) of seeing the new planet. . . . I suffer no small amount of self denial and vexation in being placed in such a climate as that of Liverpool, for I assure you I have never seen Saturn, or the new planet, or any object at about their altitude for one single minute without the annoyance of atmospheric vibration.[10]

He did, however, manage to obtain three more views of Neptune that year (1846). On November 30, he entered in his journal, 'A minute star above, and a little to the left of the continuation of the ring, distance 2 diameters. Speculum B and Merz's prism.' Evidently the ring still figured in his views of the planet, despite the poor conditions. Things improved slightly on December 3, for he then found, 'The same relative appearance exactly of planet and small star as on October 10th. The direction of the ring estimated at about 70° with the parallel of daily motion. The small star about 3 diameters distant and 50° N. following. Telescope as before.'[11]

Though Lassell observed on December 4, he did not see the ring, nor did he mention the ring; at least his record is strangely silent on the latter. Subsequent to his observation of December 3 1846, Lassell published nothing further on this matter largely because he could offer no new evidence to alter what had already been stated. That he did again see the ring is implied in the aforementioned communication to Challis, thus,

> . . . The amount of evidence of my having seen a supposed ring is very correctly recorded in the minutes of the December meeting of the Astronomical Society; for though I have again seen the same appearance, the remarkably disturbed state of the atmosphere prevented my being able to add anything to

what I had before stated. Should an opportunity occur before the planet is quite gone out of reach I will inform you of the result.[12]

This fact of more, yet unknown observations is brought out by the Rev Dawes in a letter to Challis of April 7 1847, 'In private letters to myself however he [Lassell] has given me several diagrams from time to time.'[13] (See Plate, page 135.) Whilst in a despatch to the *Astronomische Nachrichten* dated August 6 1847, Liverpool, Lassell himself said:

> With respect to the supposed ring of Neptune I am not able to add anything to what I stated last year. I have seen on two or three occasions the same appearance, but nothing more strongly confirmatory, and I await to see the planet more nearly on the meridian, and for a state of atmosphere that will bear the application of higher powers than are necessary to see the satellite.[14]

Addressing the Royal Astronomical Society at its meeting of November 12 1847, Lassell described his observations on the satellite, by then fully confirmed, but still complained that inclement weather and the low altitude of Neptune opposed his attempts to elucidate the ring phenomena. Consequently he had no alternative but to suspend all further scrutiny of it until a more adequate opportunity arose. Consequently he proposed to defer final judgement of its reality until the planet could be observed under better conditions.

As just mentioned, the other observations of October 1846, of faint stars close to the planet, had been authenticated. From the way in which the fugitive lights had kept pace with Neptune and retained the same appearance and brightness, Lassell had quickly deduced they were one and the same object, and that this was a satellite. But the situation of Neptune in relation to the Sun obliged him to delay verification until the following summer. On July 8 and 9 1847, though, he once more saw the sparkling brilliant, and rapidly established its character.[15]

If assured on this point, Lassell found little comfort as regards the ring, and as the days brought no confirmation, his occasional remarks about it only masked his emerging belief. From personal knowledge William Henry Smyth (1788–1865) was able to write in May 1847:

> The case of the ring must probably remain undecided for a few years, until the planet occupies a better situation in the ecliptic, by rising in declination to a good-working altitude: but as we happen to know personally that its existence is now doubted by Mr. Lassell himself, 'tis as well to say so.[16]

But the haunting was not over. In the Autumn of 1852 Lassell took his 24-inch telescope to Malta to observe nebulae and planetary satellites under the conditions he had long awaited to experience. Now whilst he saw the planet generally spherical in shape, and well defined under the highest powers:

> On one occasion (Nov. 4, 1852) he received a decided impression of ellipticity in the direction of the greatest elongation of the satellite, and of an extremely flattened ring in the direction of the transverse axis; but he was rather afraid of some illusion in the observations.[17]

HIND AND OTHERS OBSERVE THE RING

Lassell officially announced his discoveries to the Royal Astronomical Society on the evening of November 13 1846.[18] His paper was briefly abstracted in the published minutes of this meeting, and forms the concluding paragraph of a long list of meridian observations taken at various observatories in Europe and the United States. Since it provides additional comment, it is relevant to reproduce it here in entirety:

> Mr. Lassell has viewed the planet with his Newtonian reflector [2-feet diameter, equatorially mounted], and sees something like a ring crossing its disc. The atmospheric conditions have not been favorable, but the streak is seen in the same direction, using two different mirrors, and by several observers, so that its existence seems very probable. The appearance may possibly be caused by some distortion or flexure in the mirror, but can scarcely be the effect of imagination. One, or perhaps two, luminous points have been seen, which *may be* satellites; but this will require further scrutiny.[19]

At the next meeting of the Society held on December 11, Lassell submitted his 'Account of Physical Observations of Le Verrier's Planet', an up-dated version of his earlier paper, describing his observations up to December 4 1846 in greater detail. In effect this represented his last published paper on the question of the ring, though, as we have seen, he did mention it in later papers on the satellite.[20]

When Lassell delivered his first paper on the discovery to the Royal Astronomical Society, the Fellows present included the enthusiastic observer John Russell Hind (1823-95), then on the verge of a brilliant astronomical career. Hind had been an assistant in the Magnetic Department of the Royal Observatory, Greenwich, and had participated in the Government Chronometer Expedition to determine the longitude of Valentia. In June 1844 he had been appointed Director of George Bishop's private observatory at South Villa, Regent's Park, London, upon the recommendation of his superior, George Airy, the Astronomer Royal. Between 1844 and 1854 he discovered several new comets and minor planets, and increased the number of known variable stars. But the observation for which modern astronomy remembers him best was made, rather casually, on October 11 1852; this was T Tauri, a nebulous object afterwards found to be of variable brightness (NGC 1555). Hind wrote many papers, and two widely read books—his very popular *The Solar System* (1851) and his authoritative *The Comets* (1852). In 1853 he became Superintendent of the Nautical Almanac Office.

Hind resolved the disk of Neptune for the first time on September 30, and had subsequently observed the planet whenever atmospheric conditions allowed. The South Villa telescope, an equatorially mounted Dollond refractor of 7 inches clear aperture and 10·75 feet focal length, in all particulars an elegant and efficient instrument with a brass tube, yielded sharp images, but at no time had it given him cause to suspect the existence of a feature in any way comparable with that of which Lassell spoke. Hind decided to take a fresh look at the new planet, and on December 11 1846 reported back to the Royal Astronomical Society that in the South

Villa telescope it now assumed a decidedly oblong form, its major axis making an angle with the meridian of 30°.[21] News he had communicated to the *Astronomische Nachrichten* three days earlier:

> ... The existence of a ring appears as yet undecided, though most probable. Le Verrier [Neptune] presents an oblong appearance in Mr. Bishop's refractor.[22]

To others also, about this same period, that is, the end of 1846, Neptune appeared other than rotund. Lassell informed Challis, in his letter of January 19 1847, 'I have heard somewhat indefinitely that De Vico has observed the planet to have constantly two appendages which may be either a ring or a cluster of satellites.'[23] Francesco de Vico (1805–48) was Director of the Collegio Romano Observatory. He used a Cauchoix refractor about equivalent in size to that of the South Villa instrument.

Ormsby MacKnight Mitchel, Director of the Cincinnati Observatory, and publisher of the short-lived *Sidereal Messenger*, gives us to understand that Lt Matthew Fontaine Maury (1806–1873), Superintendent of the US Naval Observatory, Washington, DC, had seen 'points of suspected light ... *above* and *below* the planet.'[24] That he made such an observation is, however, non-proven; his literary remains are uninformative and we quote this fragment more for its interest than relevance.[25]

From Hind: *The Solar System* (1851), we transcribe the following passage:

> Professor Bond, of Cambridge, United States, who has under his direction one of the largest refracting telescopes in the world, announces that he has repeatedly seen some kind of luminous appendage to the planet, similar to what might be supposed to be the appearance of a thin flat ring, but he does not profess to say positively whether the phenomenon is really due to this cause, or whether it be owing to close satellites, which very probably exist, or to some optical illusion. It is understood that the astronomers of Pulkova, in Russia, who are in possession of an instrument precisely similar to that of Cambridge, United States, have not yet succeeded in observing any appearance such as would lead to the suspicion of a ring.[26]

north following to south preceding, or
from south following to north preceding
he does not state. — In private
letters to myself however he has given
me several diagrams from time to
time, in which the inclination of
the ring has always been drawn from
south following to _north_ preceding;

☿ , or in double star
phraseology, the angle
of position is given 160° & 340°.
If this is what Mr Lassell has observed
it is at variance with your observation

I do not at all wonder at your not
having previously seen the ring,
if the state of the atmosphere during
the autumn & early part of the winter
was no better at Cambridge than it
was here. — The nights of good

Page 135 Ring of Neptune according to the Rev W. R. Dawes from information supplied to him by Lassell. Given in a letter from Dawes to Professor Challis dated Cranbrook, 1847 April 7. Dawes directs attention to an apparent anomaly between the ring as seen and described by Challis, and as reported by Lassell. His critique stems from data supplied in correspondence by Lassell. This is at variance with the published account of the latter. See text. Original in the archives of the Cambridge Observatories.

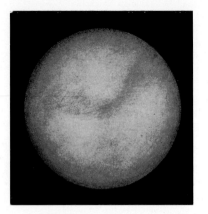 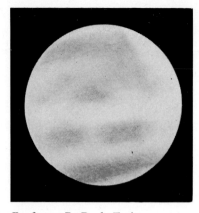

Page 136 *(left)* Neptune as drawn by Professor R. R. de Freitas Mourão, Observatorio Nacional, Rio de Janeiro, Brazil, 1959 June 30, with the 18-inch Cooke refractor. Diffuse dusky spot seen near centre of disk; *(right)* Neptune as drawn by Charles F. Capen, Jr, Planetary Research Center, Lowell Observatory, Flagstaff, USA, 1950 April 20 at 0400hr UT, with a 12·5-inch Newtonian × 432. Seeing fair. Disk mottled and blue grey in colour. South at top, in both cases.

It is, of course, true that W. C. Bond did perceive some inequality in the sphericity of the disk. His observations of Triton in the *Monthly Notices* for 1848, include a footnote on the resolving power of the great 15-inch refractor at Harvard. The final sentence reads, 'There is no appearance of a ring to Neptune when viewed with high powers, though with lower powers there seems to be an elongation.'[27] Did Hind base the foregoing on this slender fragment? Or, are we apprised of personal knowledge? Either way information is rare. The very fact that Bond disputed the existence of the ring under high magnification, but condoned its presence with a reduction, scarcely equates with what Hind tells us. In the correspondence that undoubtedly passed between them, Bond perhaps confided observations of which he published nothing. And yet, another explanation emerges. Dr Joseph Ashbrook, Editor of *Sky and Telescope*, and Dennis Rawlins, Assistant Professor of Physics and Astronomy at the College of Notre Dame of Maryland, Baltimore, both agree that here we have a possible allusion to the second spurious satellite of Neptune.[28] Rawlins in particular comments:

> I know of no other *confirmation* of the Neptune ring. But if Hind was mistaken about Bond having made one, I can at least suggest a likely explanation of his reference to such. One can easily imagine Hind recalling that Bond saw *something* new around Neptune—but getting confused about precisely *what* it was. Therefore (in the context of your inquiry) it may be more than coincidence that Wm. C. Bond was the first 'discoverer' (Lassell later, *A.N.* 31: 143-144) of the apochrypal 2nd satellite of Neptune. (Nereid was about 4 magnitudes too faint for Bond's 15 inch glass—and 3 mags., too dim for the Starfield 24 inch mirror, even assuming perfect reflectivity.) His 1847 reports (A.N. 26: 287-288 and opposite page) were rather confident: 'We have pretty strong evidence of the existence of another satellite, fainter and more distant from the Primary than Lassell's.' Lindenau later noted this report in his account of the discovery of Neptune; and I find that Edgar Allen Poe's final work (1848) *Eureka* refers to the two satellites which accompany Neptune.[29]

De Vico and Maury, it will be remembered, were unable to prescribe what they saw. Taken in conjunction with Bond's difficulty, there is some reason to believe, therefore, that these observers all saw much the same thing. James Challis, on the other hand, had no such doubts.

In his second official report on the Cambridge observations of Neptune to the Vice-Chancellor and Syndicate of that University, dated March 22 1847, Challis confidently described how he and his assistant had obtained several clear views of the ring:

A regular series of observations of the planet was commenced on Oct. 3, 1846, and continued at all available opportunities, partly with the meridian instruments, and partly with the North-umberland equatoreal, to Dec. 4, soon after which the planet became too faint to observe on the meridian on account of day-light. The observations were subsequently carried on with the equatoreal to Jan. 15. The series was much interrupted by cloudy weather, particularly in the months of December and January. . . . On Jan. 12 I had for the first time a distinct impression that the planet was surrounded by a ring. The appearance noticed was such as would be presented by a ring like that of Saturn, situated with its plane very oblique to the direction of vision. I felt convinced that the observed elongation could not be attributed to atmospheric refraction, or to any irregular action on the pencils of light, because when the object was seen most steadily I distinctly perceived a symmetrical form. My assistant, Mr. Morgan, being requested to pay particular attention to the appearance of the planet, gave the same direction of the axis of elongation as that in which it appeared to me. I saw the ring again on the evening of Jan. 14. In my note-book I remark: "The ring is very apparent with a power of 215, in a field considerably illumined by lamp-light. Its brightness seems equal to that of the planet itself." On that evening, Mr. Morgan, at my request, made a drawing of the form, which on comparison coincided very closely with a drawing made independently by myself. The ratio of the diameter of the ring to that of the planet, as measured from the drawings, is about that of 3 to 2. The angle made by the axis of the ring with a parallel of declination, in the south-preceding or north-following quarter, I estimated at 60°. By a measurement taken with the position circle on Jan. 15, under very unfavorable circumstances, this angle was found to be 65°. I am unable to account entirely for my not having noticed the ring at an earlier period of the observations. It may, however, be said that an appearance like this, which it is difficult to recognize except in a good state of the atmosphere, might for a long time escape detection, if not expressly and repeatedly looked for. To force itself on the attention, it would require to be seen under extremely favorable circumstances. Previous to the observations in January, the planet had been hid for more than three weeks by clouds. The evenings of Jan. 12 and 14 were particularly good, and the planet was at first looked at in strong twilight. Under very similar circumstances I have twice seen with the Northumberland telescope the second division of Saturn's ring.[30]

Assured that illusion had not brought the ring into existence, Challis announced his findings, accompanied by a drawing, to

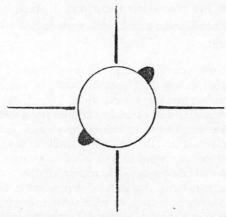

Fig 4 The ring of Neptune as delineated by Professor James Challis from observations with the Northumberland refractor, Cambridge, January 1847.

Lassell on January 16 1847[31] (Fig 4). Welcoming this news, Lassell replied on the 19th:

I cannot refuse to consider that your observation puts beyond reasonable doubt the reality of mine; especially as even your measured angle of position agrees with my estimation within 4 degrees.[32]

Challis concurred, and in identical communications to the Royal Astronomical Society and Schumacher's *Astronomische*

Nachrichten, ventured to express his conviction of the existence of the ring, thus:

> I have been able with the Northumberland Telescope to verify Mr. Lassell's suspicion that the planet has a Ring. I first received the impression of a Ring on Jan. 12. . . . The angle made by the diameter of the Ring with a parallel of declination was 66° by a not very satisfactory measurement taken on Jan. 15. The ratio of the diameter of the Ring to that of the Planet is by estimation that of 3 to 2. I am unable to account for my not having noticed the Ring earlier.[33]

But Dawes interposed some conflicting evidence. Writing to Challis from Cranbrook, April 7 1847, he pointed out that according to data in his possession there was, in fact, a serious discrepancy between Challis and Lassell with regard to the orientation of the ring:

> I have to thank you for a copy of your "Second Report of Proceedings respecting the Planet *Neptune*": as well as of your "First Report" on the same subject which I fear I was prevented from acknowledging at the time by the pressure of indisposition.
> Your proceedings are very interesting to me:– but I cannot quite comprehend about the *ring*. Mr. Lassell has not I think distinctly mentioned in any of his published statements respecting it, *in what direction the ring lies*.– He merely says that it is "about 70° from the parallel of declination" but whether from north following to south preceding, or from south following to north preceding he does not state. In private letters to myself however he has given me several diagrams from time to time, in which the inclination of the ring has always been drawn from *south* following to *north* preceding; or in double star phraseology, the angle of position is given 160° & 340°. If this is what Mr. Lassell has observed it is at variance with your observation.[34]

Dawes penned a small representation of Neptune and the ring to illustrate his point. But this *does not* correspond with that given by Lassell to Challis on January 19. As the originals are missing it is fruitless to pursue the matter further. Dawes must be disregarded in favour of Lassell, yet his remarks must apply with qualification in any discussion on the subject.

By this time interest in the ring had reached its climax. In its report to the 27th Annual General Meeting, the Council of the Royal Astronomical Society declared on February 12 1847:

> Mr. Lassell, of Liverpool, conceives that he has seen a ring, and probably a satellite, attached to the planet. The existence of the ring seems almost certain, as other observers, with far inferior telescopes, have noted some irregularity in the round-ness of the planet, and Professor Challis, with the Northumber-land telescope, nearly coincides with Mr. Lassell in his des-cription of the appearance.[35]

Assailed by doubt, others took a more cynical view, and that summer Lassell himself felt the thrust of emergent disquiet; a feeling that became more insistent with each successive opposition of the planet, until some five years later, under the immaculate skies of Malta, though he again suspected the ghost of a ring just once more, his confidence by then was so completely undermined that he denounced the appearance instantly, and abandoned it to illusion. And so the ring of Neptune vanished in the penetrating light of modern researches. Conspicuous at discovery, it has singularly failed to re-present itself at any time, and whilst briefly recalled by Harlow Shapley in September 1950, when he specu-lated on the possible existence of other ring structures in the planetary system and explored ways and means of detecting them, it has long languished in the obscurity that is the common heritage of all such dubious features.[36]

CONCLUDING REMARKS

Any attempt to deduce the nature of the appearance that deceived so many astronomers into believing they had seen Neptune encircled by a ring structure, is limited by a critical lack of positive information, and any explanation is, therefore, neces-sarily tentative and conjectural. At the beginning we drew a dis-tinction between the ring of Neptune and that of Uranus. This is justified on the grounds that with the latter, the explanation is reasonably clear. The Front-view arrangement was in its infancy, and subject in consequence to the teething troubles that beset all new contrivances. To endow the ring of Neptune in the same manner would merit attention only in respect of Lassell's observations. It would singularly fail to account for the reports of Bond, De Vico, Challis, Hind, and Maury, since they used refracting telescopes. And whilst astigmatism applies equally in this direction, it would we must agree be a most remarkable coincidence that so many telescopes at one and the same time were all similarly affected by this defect. If we admit fault in Lassell's 24-inch, we must look elsewhere in order to explain all the other observations.

Let us review the evidence in summary. In the winter of 1846-7, Neptune was found to be accompanied by a thin, flat, elliptical form of high eccentricity, which when most clearly seen gave the undoubted impression of a ring obliquely inclined to the line of vision. This feature was neither clear nor obvious, but tenuous and faint, and required reasonable conditions for its detection. It was only definitely seen in large apertures. Thus in a 24-inch reflector it assumed a linear form, as also in the 11·75-inch Northumberland refractor. In the 15-inch refractor of the Western Cambridge, low magnification revealed an elongation to the disk which disappeared when viewed with greater power. Small apertures gave only uncertain impressions of abnormality. It is significant to relate the experience of the Rev W. R. Dawes:

> Only on one occasion did I see it (Neptune) in a favorable state of the air; then it appeared with power 195 to be sharply defined and *circular*, having no projection whatever.[37]

Dawes used a fine 6⅓-inch refractor.

We must also repeat that if the ring was discovered by a reflecting telescope, confirmation was exclusively the province of the refractor. To subdue any charge of malfunction in his optical system, Lassell employed two different mirrors, and two different secondaries (a plane mirror and a Merz prism).

The observers involved in the case, Bond, Challis, De Vico, Hind, Lassell and Maury, were accomplished and fully experienced, and would we anticipate be well able to discriminate between the false and the real.

We gather from Challis and Lassell that the weather in England during the late autumn and winter of 1846 was incredibly bad, there being an undue prevalence of cloud. Dawes confirms this in his letter of April 7 1847:

> I do not at all wonder at your not having previously seen the ring, if the state of the atmosphere during the autumn and early part of the winter was no better at Cambridge than it was here.[38]

This only partially explains Challis's puzzlement over not having 'noticed the ring at an earlier period.'[39] At that time, however, he was preoccupied with positional work; taking meridian observations with small aperture equipment. Hind with the 7-inch South Villa refractor could only make out a slight ellipticity in the shape of Neptune, which he considered due to the ring. It is not surprising, therefore, that Challis failed to see anything unusual until he switched to a larger telescope.

Against this setting two explanations may be framed:

(i) That Neptune is encircled by a ring of such tenuity that it escapes common detection and might be fogged out on a photographic plate.

or (ii) That optical defect plus pre-suggestion precipitated the whole business.

Six sets of observations are available:

(i) The detailed accounts by Lassell and Challis.

and (ii) The vague circumstantial evidence of Bond, De Vico, Hind, and Maury.

If we accept the Ashbrook-Rawlins hypothesis then we can reject Bond's observation, which is reasonable since he himself emphasised that the elongation effect was evident only in low powers and disappeared when the planet was examined with greater magnification. The data supplied by De Vico, Hind, and Maury is insubstantial and can be disregarded. This leaves the Lassell-Challis series, but can this be trusted?

Take Lassell's observations. Remember Neptune was badly placed for observation from the northern hemisphere, and that atmospheric conditions were very disturbed. Lassell states the ring was independently described by several unknown persons using the 24-inch. But who were these observers? Most probably they were merely visitors to Starfield who had the privilege of viewing the newly discovered planet through a very large telescope. Like the elder Herschel, Lassell introduced many optical and mechanical innovations into his construction of telescopes, of which the most relevant in this connection was his adoption of a rectangular glass prism as a secondary mirror or flat. He used two speculum metal mirrors; Speculum A used in conjunction with a plane flat, and Speculum B with the prism. The discovery was made with Speculum A. The following extract from Lassell's letter to Challis of January 19 1847 is of some interest, not for what it says, but for what it implies of the condition of Speculum A during the critical period:

> I have pleasure in being able to add that I have within the last three weeks put a new surface upon my two foot speculum A which appears to me to be absolutely, or very nearly perfect. I have it is true only had a portion of one evening when the sky cleared off for a short time after 10 o'clock, on which to try its performance; but I am nevertheless well satisfied from the various tests employed that it will take much time to discover either the amount or direction of the error which may exist.[40]

Would it be correct to deduce that at the time when Lassell saw the ring, his Speculum A was slightly in need of maintenance? That he observed a remnant of the ring from Malta some five years later, with the same instrument, does suggest an inherent fault; yet if so, it is remarkable it did not manifest itself on other objects, such as faint planetary satellites.

On the other hand, is Challis to be regarded as an independent witness? Was he so unbiased in his reception of the appearance? We think not. By January 1847 Lassell's observations had been published in full in the *Monthly Notices*, from whence therefore Challis obtained his information. Little doubt Morgan also read these reports. Hence when Challis asked Morgan to pay especial regard to the physical aspect of the disk of Neptune those nights in January 1847, what more natural than his thoughts should turn instantly to the very latest observations, namely those of Lassell. Surely this could have affected his view of the planet. Imagination and pre-suggestion are powerful forces to reckon with when observing features at or near the threshold of visibility; one only has to refer to the observational history of Mars to appreciate this most important fact.

In conclusion, let us attempt to reconstruct the circumstances of how things might have been. We can well imagine the optimism that prevailed during the summer of 1846, after the publication of Le Verrier's prediction. At a special meeting of the Board of Visitors of the Greenwich Observatory, on June 29, George Airy casually remarked the likelihood of a new primary planet. It made a profound impression; on September 10 Sir John Herschel, who had been present, confidently announced to the British Association for the Advancement of Science, then assembled at Southampton, that:

> We see it [the new planet] as Columbus saw America from the shores of Spain. Its movements have been felt trembling along the far-reaching line of our analysis with a certainty hardly inferior to that of ocular demonstration.[41]

How did Lassell react when news of the momentous discovery broke late that same month? It entails no great stretch of the imagination. At the earliest opportunity he doubtless swung the great 24-inch in the direction assigned, among the first to encompass the breadth of the virgin world. We picture him intent on its grey-green disk, dimmed and reduced with distance, vibrating unceasingly in the disturbed air. He becomes aware of something strange about its appearance; something he can scarcely comprehend. Perhaps a close satellite or two? Or a near configuration of faint telescopic stars? Illusion perhaps? No matter what the explanation, in poor seeing, with the planet badly placed for observation, the impression crystallises into a definite shape, which in his effort to seek a distinctive trait to the planet, Lassell interprets as a planetary ring. Having seen it once, the impression remains sustained by imagination. Visitors to the observatory are asked to look, and they confirm the initial impression. But the probability is that these attestors were not trained astronomers and so would look with an uncritical eye. Thus, his suspicions confirmed, Lassell makes a short announcement of his discovery. Reputation and telescope held in high esteem, the report is dignified and acknowledged in the published minutes of a learned society. Conditioned to expect others to search for and claim 'an impression of a ring', or something that could well be a ring. Yet previously they suspected nothing untoward about the planet. Carried abroad, the intelligence, perhaps transposed in translation, affects others and further reports filter through, but in the end, the matter so precarious from the start, is doubted and thought to have been founded on illusion.

From our sophisticated standpoint, these small episodes from the observational record of Uranus and Neptune are curious, but they do not reflect discredit on the participants. In some ways the nineteenth century was the golden age of planetary discovery. Everything new and largely unknown, so that even the smallest telescope ventured out on to strange uncharted seas on voyages akin to those of classical mythology. Mistakes were made and legend proliferated. In the context of the day such observations appear less fanciful than at first imagined; as Alexander von Humboldt wrote of the lost satellite of Venus, they 'belong to the astronomical myths of an uncritical age.'[42]

7 Strange Interlude

> *From the fact that Neptune is the farthest planet we know we have by no means the right to conclude that there are not others beyond it.*
>
> Camille Flammarion (1894)

DISCOVERY and exploration are potentially rich sources of myth and legend. A fact as true in astronomy as in the more domestic environment. Nowhere is it more evident than in the history of planetary discovery. For as in bygone days rumours in a distant market place sent the restless in search of hidden cities and lost continents, so in astronomy men have swept the heavens for planets of dubious existence.

Late in the nineteenth century, Niesten of the Royal Observatory, Brussels, created a minor sensation when he proposed that the apocryphal satellite of Venus, reported by Cassini, Short, Montaigne and others during the previous two centuries, was not in reality a satellite at all, but a tiny planet circling the Sun between Venus and the Earth. A body he provisionally named Neith, after the Egyptian Minerva, hidden behind her veil. That same century astronomers were keeping a close watch on the Sun in the hope of detecting some trace of the mythical Vulcan, guided by the clear and definite principle of the intramercurial hypothesis of Le Verrier. Equally cogent reasons directed others into the outer regions on the track of the hypothetical planets of W. H. Pickering, but again without success. Equally strange is the story of Themis, the enigmatic moon of Saturn, found then lost. The catalogue of such incidents is long and fascinating, and like its terrestrial analogy, stirs the imagination with a prospect of part of the unfamiliar debris which borders the path cut by modern astronomy into the unknown. It is a bizarre history of quixotic notions, of error, the explained and the unexplained. In short, an extraordinary adjunct to the mainstream of astronomical progress.

In this and our final chapter, we propose to run through the events of three alleged discoveries that took place during the grand sweep of discovery that disclosed the existence of the outer planets. We point no moral, and establish no thesis, but merely present the details as they are known, as examples of peculiar observations. The first of these reports has three points of immediate interest. First, it anticipated the discovery of Pluto by some eighty years. Second, it is one of the few episodes in this genre ever fully explained. Third, and last, it may be regarded as a cautionary tale of some moment. The lost planets to which we refer in chapter 6 have never been accounted for, assuming the authenticity of the observations.

HYGEIA OBSERVED

On April 12 1849, Annibale de Gasparis (1819–92), Director of the Capodimonte Observatory, Naples, discovered the minor planet Hygeia, the tenth such body to be found.[1] Just over a year later James Ferguson, astronomer on the staff of the United States Naval Observatory, Washington, DC (then the National Observatory), began a routine series of observations on its motion. Using the large Equatorial, a 9·6-inch refractor by Merz and Mahler of Munich, installed in 1844, Ferguson spent the summer and autumn months of 1850 tracking Hygeia as it cruised through the rich star fields of Sagittarius, determining its position by reference to adjacent stars of known location by means of a filar micrometer.

This is a elaborate piece of apparatus mounted in the focal plane of an equatorial refractor and used to measure small angular distances on the celestial sphere. First described by Auzout in 1667 in a form not appreciably different from that of today, it consists basically of a flat, rectangular box across which two fine wires are stretched. These are fixed and at right angles to each other. Another wire parallel to one of the former is drawn taut across a frame free to slide within this box. The movement of the frame is controlled by a finely pitched screw connected to a graduated drum, from which the observer can read off the total of revolutions applied to the screw. Important to the case in hand, the Washington micrometer contained a movable metal plate bearing *three east-west wires*, denoted 1, 2 and 3. This served to measure the angular distance (declination difference) between two objects, one of known position, the other not. To determine the position of the latter, the observer shifted the metal plate to obtain a bisection of the known or comparison object (usually a catalogue star) by one of the three wires. The other object was then bisected by a fourth but adjustable wire also in the east-west direction. The revolutions performed by the screw thus provided a measure of the angular distance traversed by the wires when the angular value of one revolution was known. Hence from the settings of the plate and the movable wire the declination difference could be worked out, and the position of the unknown body derived by computation.[2]

Ferguson thus chased Hygeia from May 18 to November 24 1850, fixing its position in the above manner on sixty-three nights, and describing it as '... very minute (12.13 mag.,) ... well seen,' on November 21.[3] A report was submitted to Lt M. F. Maury, Superintendent of the Observatory, who passed it for publication to *The Astronomical Journal*. Entitled 'Observations of Hygea made with the Filar-Micrometer of the Washington Equatoreal' it appeared in print 1851 January 18, and so came before John Russell Hind, the well known English observer of whom we spoke in our last essay.

'K'

In a sense Hind had a vested interest in this report. By sheer coincidence he had been patrolling the same territory as Ferguson, but for entirely different reasons; '... for the past four years.' as he later wrote, 'I have constantly had this very region of the heavens under close examination, having in August 1847, missed a small star in about R.A. 19h 1m 45s, N.P.D. 111° 56' (1800).'[4] Furthermore, a devoted student of the minor planets, already with three discoveries to his credit and to add seven more to this total

by July 1854, he had also been charting all stars down to the
eleventh magnitude situated 3° on either side of the ecliptic, to
facilitate future discoveries of these bodies. The intention being
to produce a complete map of the zodiacal stars, divided into
twenty-four zones, one for each hour of right ascension. In the
period concerned Hind had been '. . . filling in all stars . . . for
Hour XIX of our charts.'[5] Curiously, precisely the region navi-
gated by Hygeia in October 1850.

It is important to remark at this point, that of the stars used
by Ferguson to deduce the position of Hygeia, many were un-
catalogued. Hence in advance of his purpose of employing them
as comparisons, he first micrometrically determined their co-
ordinates by reference to nearby catalogue stars, and identified
them with a temporary local prefix such as a letter or numeral. It
was natural, therefore, in those days of incomplete stellar cartography, for one observer to check his results against those of
another in matters of this nature.

Now, in the course of his comparison, Hind noted that on the
night of October 21 1850, Hygeia had been referred to a 9.10
magnitude star, listed as 'k'. From three observations with the
West Equatorial based on star 1719 of the *Greenwich Twelve Year
Catalogue*, Ferguson quoted the adopted mean place of 'k' for
1850.0 as RA 19h 17m 40s·60 and Dec −20° 45' 05"·68.[6]

This puzzled Hind. He failed to recollect '. . . any suspicious
body in the vicinity.'[7] More peculiar, 'k' could not be reconciled
with any star on his chart, nor yet could it be found when he came
to search for it in the heavens. Hind reflected. Had Ferguson
made a mistake, either in his observations or reductions? How
could he blatantly misplace so obvious an object as a ninth
magnitude star? Had he, in fact, substituted one star for another?
Hind well knew the problem of identification in thickly populated
regions, such as the planet had traversed in October 1850.
Ferguson himself had remarked: 'From this date, August 26, till
the termination of the series, there were many small stars about
the path of the planet, making it necessary to observe more than
one star. Occasionally the wrong star was observed, and the

observation lost.'⁸ This actually happened on September 20 1850
—'. . . the wrong star observed. The single comparison given was
accidental.'⁹ Even so, Hind reasoned, this line of thought had
little to commend it. In general Ferguson had shown himself to
be a competent observer. Other than errors due to personal
equation, his record seemed entirely trustworthy. Anyhow, the
fact that the star could not be found did not necessarily imply
error. On the contrary, Hind argued, at least two other explanations were evident. Either there was an unknown variable star
in the case, or else Ferguson had accidentally discovered a new
minor planet. For obvious reasons, Hind was inclined to favour
the second hypothesis, and in a letter to W. C. Bond, Director
of the Harvard College Astronomical Observatory, not only did
he report the missing star, but called attention to its possible
planetary character. It was all very circumstantial and uncertain,
yet Bond was intrigued enough to inform Lt Maury at Washing-
ton.¹⁰

FERGUSON VERIFIES 'K' HAS DISAPPEARED

Maury's immediate reaction involved Ferguson in a laborious
session with the large refractor. Instructed '. . . to examine the
proper part of the heavens for the star in question,' he diligently
checked the suspect zone on the night of August 29 1851, only
to report the star absent.¹¹

This circumstance induced Maury to audit the relevant entries
in the observing journals for October 1850.¹² As it turned out
Ferguson had first seen the star on October 16 of that year, when
he was looking for Hygeia, and had compared it with Greenwich
star 1719. It had again been sighted on the 19th, with two unknown stars; observed on the 21st, with Hygeia, though with
regard to the latter Ferguson had remarked 'Doubtful if planet,
and very badly seen.'¹³ And finally on the 22nd, it had been compared with Lalande's ninth magnitude star 36878. Throughout, the
Washington astronomer had designated the object as a star of
9.10 magnitude.

Significantly, these observations indicated that the object had motion in right ascension, there being strong grounds to believe it had been stationary at some time between October 16 and 22 1850. Suspicion undoubtedly attached to the object, and Maury had little alternative but to concur with Hind. Accordingly, on September 3 1851, in a letter addressed to William A. Graham, Honorary Secretary of the US Navy, Maury reported that:

> The star of comparison with Hygea on the night of October 21, 1850, has disappeared. It is not now to be found where it then was. Hence I infer that it is an unknown planet.

Believing his Observatory had missed a rare opportunity, Maury concluded on a regretful note:

> . . . and that those observations, had there been sufficient force at the Observatory for their immediate reduction, would then have revealed to us the character of this star.[14]

To this *communiqué* were appended the observed places of 'k' for October 16 and 22 1850, as follows,

Date 1850	Right Ascension			Declination		
	h	m	s	°	′	″
Oct 16	19	17	40·605	−20	45	6·58
Oct 22	19	17	42·364	−20	45	3·42

HIND MAKES AN ANNOUNCEMENT

That summer (1851) Hind and De Gasparis both logged their fourth discoveries of minor planets. Irene was found by the former on May 19, and Eunomia on July 29 by the latter. Quite possibly these events stimulated Hind's interest in the truant star. In the August he discussed the affair with a visiting American astronomer Benjamin Apthorp Gould (1824–96), founder and editor of *The Astronomical Journal*, and ventured to express his opinion that it would not be explained '. . . on any other supposition than by assuming it to have been an unknown planet.'[15] By which he of course meant a minor planet, as probably intimated in his earlier letter to W. C. Bond. Lt Maury certainly believed this to be so, and after the abortive sweep of August 29, he encouraged Ferguson to continue the hunt upon this premise.[16] But the catch was of greater import, or so it seemed.

Hind realised this after reading Maury's letter of September 3, which had been published in *The Astronomical Journal* of October 22 1851, together with the observed places of the star on October 16 and 22 1850. The motion of the star as indicated by these observations Hind noted, was inconsistent with his original belief. Instead it suggested something vastly more exciting. Something he averred '. . . I conceive it highly desirable . . . should be in the hands of astronomers.'[17]

As yet Maury had deferred any public announcement of his search. Hind, however, knew of its progress from Gould: 'You tell me', as he wrote to the latter on November 12, 'that the Washington observers are engaged in watching the small stars about the place of the one missing, but their silence up to the present date is discouraging.'[18] Other than what is implied in this quotation, no positive authority exists to establish the extent of this correspondence. That it took place is self-evident. In other details also we can only infer what transpired between Hind and Maury with Gould acting as intermediary. Further, we also suspect that Hind was conducting his own search for the supposed planet, but again time has obscured the details. Be this as it may, Hind seized the opportunity of this exchange to acquaint Gould of the real significance of the Washington observations. On November 12 1851 Hind wrote the letter which, for a time at least, set the wires of the astronomical exchange vibrating with the most extraordinary intelligence. Recalling their conversation in the August, Hind continued:

In number 31 of your excellent journal Lieutenant MAURY has announced that the original observations have been examined, and that farther evidence in favor of the planetary nature of the object has been adduced, the positions resulting from the differential observations of October 16 and 22d showing an increase in right-ascension on the latter day of $1^s.76$, which is quite beyond any probable error of observation on the part of a skilful observer, such as Mr. FERGUSON has proved himself to be. Now a mere glance at the places of the suspicious object, and of the sun at the time, will be sufficient to convince us, that, if there be a planet in the case, it could not have belonged to the prolific group between *Mars* and *Jupiter*, that its mean distance in fact must be greater than that of any known planet . . . If we assume (as I think we may safely do, on the further hypothesis of circular motion), that Mr. FERGUSON's star would be *stationary* within a day or two before or after October 16, we shall find that a planet in this position must have a distance of more than 137 [AU's], and a period of above 1600 years.[19]

Startling in his forecast, Hind merely amplified a growing conviction in certain quarters. The old concept of the Solar System as a compact, finite unit following a pre-ordained path through space had been condemned in 1781 by the discovery of Uranus, and banished by Giuseppe Piazzi when he found a strange seventh magnitude object amongst the stars of Taurus on the night of January 1 1801, and so initiated knowledge of the minor planets. Authentication of the trans-Uranian hypothesis in September 1846 led to further speculation. Hardly a week had passed since that event, than Le Verrier confided to a colleague that in his opinion Neptune did not mark the frontier of our planetary system. The fact that the gravitational field of the Sun extended beyond Neptune made it certain that other planets might yet be found in those remote regions. Thus Hind projected a view already current, and in the case of 'k' transmuted speculation into hard fact.

Even so, Hind still felt apprehensive. The influence of suppressed doubts was apparent as he further remarked to Gould on November 12 1851:

. . . I am at a loss to imagine how a slow-moving planet of 9.10 magnitude can have escaped me. Such a planet would be easily recovered, and I am convinced I must have seen it during the past summer, if it retained the same degree of brilliancy assigned by FERGUSON (9.10 magnitude). So far, then, as my own search has extended, I feel able to state confidently that there is no planet hereabouts of the 9.10 magnitude.[19]

Further:

> ... The *second* observation on October 21 is not given in your
> journal; perhaps Lieutenant MAURY may favor us with the
> *original comparisons* by which the star's places are determined.[19]

Yet, Hind conjectured, just suppose:

> ... that *its light is subject to variation* so as to cause it to descend to
> the eleventh class, then I should think it necessary to institute a
> further examination before pronouncing an opinion upon the
> subject.

THE WASHINGTON SEARCH

Gould published this letter in his journal of January 1 1852, but
had apprised Maury of its content some time before. Consequently
the search for 'k' was re-orientated '. . . upon the supposition that
it is not an asteroid, but an exterior planet.'[20]

A year had now elapsed since the fugitive had last been sighted.
Ferguson doubtless faced the prospect of its recovery with some
disquiet, justifiably so in the circumstances. Photographic search
methods were non-existent, whilst available cartography gave
scant coverage of the region concerned; the great *Bonner Dürchmusterung* did not appear until 1863. If indeed a planet was abroad
as Hind and Maury supposed, then clearly its presumed motion
daily carried it deeper into the dense and largely unexplored fields
of Sagittarius, increasing the difficulties already in the way of its
recovery.

Still the search had begun, optimistically enough, on the night
of August 29 1851 with a general sweep of the suspect area.
Spurred on by the thought of discovery, Ferguson spent four
tedious months at the ocular of the large refractor checking and
re-checking a bewildering array of faint and unknown stars in the
hope of detecting some trace of movement that would betray
his planet, but to no avail. Quite inexplicably 'k' had vanished.
Realising the futility of prolonging the search, Maury decided
upon its termination on December 11. In the interim Ferguson
had examined all stars down to the eleventh magnitude, as recom-
mended by Hind, between RA 19h 20m and 19h 36m, and Dec

−19° to − 21° 20′. Reporting to Gould on January 5 1852, Maury baldly observed:

> Since my communication of the 3d September, 1851, relative to the missing star of October, 1850, I have had a thorough examination made of the part of the heavens in which it was seen, but without its rediscovery.[21]

Details of the observations for October 1850 were quoted, as Table on page 157.

From which the apparent places of 'k', corrected for refraction, on October 16, 21 and 22 had been deduced, thus:

Apparent Places of 'k'

		α	δ
	h m s	h m s	
Oct 16	6 52 36.3	19 17 42.81	−20° 44′ 57″.096
21	7 6 39.9	19 17 42.19	20 44 55.53
22	6 35 35.1	19 17 43.90	−20 44 54.642

As accurate positions for the stars of comparison were not available, the place of 'k' on October 19 was omitted.[22]

And there the matter rested. Somehow, the Hind-Ferguson object had eluded its pursuers. Indistinguishable from the star-studded backdrop, it had slipped unnoticed into the unrelieved gloom beyond Neptune, or so it appeared. The object had vanished, but it had left a legacy of doubt. No one had been utterly convinced of its reality. In fact, did the neatly canonised figures of the Washington statistics for 1850 really allude to the spoor of an unknown planet? Or were the observations that induced the speculation affected by errors of right ascension much greater than at first appreciated?

The mystery was to remain unsolved and all but forgotten for practically three decades. And of the principals involved only Hind and Gould, by then astronomers of repute, survived to learn the truth about those puzzling measures.

Observations of the Missing Star 'k', October, 1850

Date 1850	M.T. Washington h m s	No. of Comp	Comparison-Star	Δα m s	k — * Δδ	A	Remarks
Oct. 16	6 52 36.3	2	Gr. 12Y. C. 1719	+3 53.70	+10′ 00″61	6	This star observed for Hygea.
19	7 30 28.5	5	o	−1 21.80	+7 18.56	7	
19	7 27 28.5	1	p	−4 57.74	−9 09.31	7	
21	7 06 39.9	3	Hygea.	−3 36.91	+4 57.31	7	Doubtful if planet, and very
22	6 35 35.1	1	Lalande 36878	−5 33.47	−2 21.68	8	badly seen.

THE RECORD OF THE WIRES

The solution unexpectedly turned up at the Naval Observatory, Washington DC one morning in December 1878. It took the form of a letter addressed to the Superintendent, Rear-Admiral John Rodgers. Dated December 2, and postmarked New York, it originated from Professor C. H. F. Peters (1813–90), the German-born Director of the Litchfield Observatory, Hamilton College, Clinton, New York. Astronomer, mathematician and ex-soldier of fortune, having served in Turkey, and fought under Garibaldi in Sicily and Italy, Peters had settled in the United States in 1854 as a political exile, and some four years later had become Director of the Litchfield Observatory.

We are not aware of his motives, but it is enough to state that Peters was an inveterate hunter of minor planets, and was wholly absorbed by the question of unknown planets. He found his first minor planet, Feronia, May 29, 1861 and at the time now considered had thirty-two discoveries to his credit. Like Hind, he also had devised a scheme to chart the ecliptical stars, the intention being to produce a set of very accurate maps showing all stars down to the 14th magnitude.

New, missing and hypothetical planets had been popular topics of discussion throughout 1878. There had been twelve minor planet catches, five of which were due to Peters. After a lapse of several years, the Vulcan controversy had been reopened. During the total solar eclipse of July 28 of that year, two American observers, Professor James C. Watson, of Ann Arbor, and Lewis Swift, of the Warner Observatory, each claimed to have sighted close to the Sun small disk-like objects which they considered to be planets inside the orbit of Mercury. Now in reporting these and other eclipse studies in a *communiqué* from Washington DC, dated August 8, published in *Nature* (August 29 1878), J. Norman Lockyer, the well known English solar physicist, casually drew attention to other developments of a speculative nature by remarking 'Prof. Watson, of Ann Arbor, . . . broke off work on a planet beyond Neptune . . .'[23] The reference was timely.

As we have already said, Le Verrier had speculated on such a possibility shortly after the discovery of Neptune. Now others were taking up the challenge. On thirty clear, moonless nights between November 3 1877 and March 5 1878, D. P. Todd, assistant at the Nautical Almanac Office, Washington DC, had conducted a limited telescopic search for a transneptunian planet. Whilst across the Atlantic, Professor G. Forbes, of Edinburgh, and Camille Flammarion in Paris, were investigating the theory of how such a body might reveal itself.

It was thus merely a matter of time before someone would remember the runaway object of 1850. Its history was briefly recalled in *Nature* (October 31 1878) by an anonymous writer who quoted the observed places, and framed the enigma within the uncertain arguments adopted by Hind in 1851.[24]

Peters had been far from impressed by the Swift-Watson allegations of intramercurial planets, and after reading the various accounts of their observations had decided to analyse all the available data. In short, Peters thought the whole concept arose from errors in Le Verrier's 1859 calculations, and that the field reports related to anything but inner planets. He was thus engrossed when he read of the Hind-Ferguson episode in *Nature*.

Peters frowned on the idea that an exterior planet had been revealed. To him the residuals in the positional measures suggested error, and nothing else. He looked up the published correspondence, and checked the official volume of observations for 1850. After some mathematical detective work based on the description of the Washington micrometer, he immediately suspected what had gone wrong in October 1850. It was the result of this research that prompted the letter of December 2; obviously his conclusions would have to be verified against the original manuscript entries, and this could only be done at the Naval Observatory. Prefacing his result with a short history of the affair, Peters proceeded to give his explanation:

In order, now, that nobody thereby might be induced to spend months and years upon a renewed search, I hasten to bring to your knowledge the errors I have detected in the "Washington Observations for 1850" (on pages 320 & 321), and which alone have given rise to the misconception. This pseudo-planet indeed is nothing but a star observed already by Lalande, occurring besides three times in Argelander's Zones and twice in Lamont's,— as I am now going to show.

I will first put together the places of the stars that will come into consideration, denoting them for simplicity of reference with letters.

Star	Mean AR. 1850.0	Mean Dec. 1850.0	Authority
(a)	19h 13m 47s.48	$-20°55'$ $5''.7$	Oe. Arg. 19444-6, Lam. 886, Gr. 12y. Cat. 1719, Yarnall 8268.
(b)	17 40 .96	52 50 .6	LL. 36613, Oe. -Arg. 19535-7, Lam. 903.
(c)	19 2 .99	60 7 .5	Washington Equ. Obs. 1852 pag. 448, and 1861 pag. 391.
(d)	22 39 .24	43 47 .8	Yarnall 8343.
(e)	19 23 15 .82	-20 42 45 .1	LL. 36878, Oe.-Arg. 19669, Lam. 930, Yarnall 8350.

The movable plate of the micrometer, that served for measuring the differences in declination, carried three wires, denoted by 1, 2, 3 (Wash. Obs. 1850 Introc. page XXX). The object in question, marked (* k), was compared on Oct. 16 twice with the star (a), on Oct. 19, 5 times with (c), on Oct. 21, 3 times with the planet Hygea, on Oct. 22 once with star (e). Now it happens that for all the 11 instances when (* k) was the object of pointing, the wire was written wrongly as Nr. 1, while it was in fact Nr. 2. Making this correction, 2 instead of 1 under the heading "Mic. w." in column 7 wherever (* k) occurs on pp. 320–321 of "Obs. 1850", the 9th column, with regard to the wire distances given in the Introduction will stand thus:

Oct. 16 for	$+39^{rev}217$	read	$+9^{rev}.050$	
„	39.017	„	08.851	
Oct. 19 „	$+58.451$	„	$+28.283$	
„	58.513	„	28.345	
„	58.615	„	28.447	
„	58.600	„	28.432	
„	58.722	„	28.554	

Oct. 21	„	−19.369	„	+10.798
	„	19.405	„	10.767
	„	19.267	„	10.900
Oct. 22	„	−09.207	„	−39.374[25]

Although substitution of wire No 2 for wire No 1 only rectified the anomalies in declination, those in right ascension were automatically reduced since the tilt of the north-south wires was now properly described. Taking the daily means of the screw values, as listed, converting them into arc, and also transcribing the differences in right ascension from the printed volume of observations, Peters tabulated his results for the comparisons with (* k) on October 16, 19 and 22, stars (a), (c) and (e) respectively. Having applied these corrections, Peters at last revealed the true identity of (* k). From the observations on the three nights specified he determined its mean place for 1850·0 as RA 19h 17m 41s·09 and Dec −20° 52′ 49″·5. And this, as he then disclosed, coincided with a known star of no great significance—Lalande 36613 (b in the list given by Peters)! Ignominiously, the Hind–Ferguson transneptunian planet had been downgraded to the status of an ordinary catalogue star.[26]

After correcting a minor error in the second set of measures of October 19 1850, and by computation confirmed that Ferguson did indeed observe Hygeia on the night of October 21, despite his doubts to the contrary, Peters concluded: 'In order to stop any further perpetuation of the credence, that a trans-Neptunian planet is revealed by the Washington Observations discussed, it would be very desirable to give publicity to the corrections indicated.'[27]

Official response was immediate. Rear-Admiral Rodgers despatched the paper to the *Astronomische Nachrichten*, and in his covering letter of December 6 1878 simplified its technical character:

The explanation given by Professor Peters is entirely satisfactory. This explanation is put beyond doubt by the fact that Professor Hall finds, on examining the original observing-books, that Mr. Ferguson actually observed the difference of declination correctly on every occasion, except on that of the two transits of Oct. 16

and the first transit of Oct. 19.—At these transits, the wire was recorded 1, but at all the other transits, on Oct. 19 as well as on Oct. 21 and Oct. 22, the wire was recorded 2. For some unknown reason, Mr. Ferguson in his reductions changed all his correct observations to correspond with the erroneous ones.[28]

Convincingly, its masquerade as a transneptunian planet had been exposed, and 'k' retreated into obscurity. And yet its little jest was to be remembered in quite a different manner. For in 1931 it again turned up, but as entry 1689 in E. Zinner's catalogue of suspected variable stars! A curious end to a remarkable story.[29]

The Hind-Ferguson episode is rather unique. It is the history of a planet found by accident, but lost in embarrassment; a remarkable compound of error and prejudice set in a period when the discovery of planets was comparatively novel in terms of human experience. Ferguson may stand shadowed by his curious aberration; and Hind pilloried for his incautious intervention. Yet each embodied the spirit of the day. A spirit truly immortalised by John Keats:

> Then felt I like some watcher of the skies
> When a new planet swims into his ken.[30]

8 The Lost Planets
of 1831 and 1835

> . . . *we shall still go on finding more*
> *planets, till space, in our solar system*
> *at least, may appear, in reference to*
> *what was formerly known, comparatively*
> *full of these travelling bodies.*
>
> George Biddell Airy (1851)

In November 1758, the French mathematician Alexis Claude Clairaut (1713–65) completed his investigation into the theory of planetary perturbations in connection with the impending return of Halley's comet, and announced his results to the French Academy of Sciences. Working with Lalande and a Madame Lepaute, he had rigorously calculated the probable effects produced on the motion of the comet by the disturbing influence of Jupiter and Saturn, and had concluded that in consequence of such action the comet's return would be delayed by the attraction of Jupiter and Saturn, no less than 518 and 100 days respectively. To his prediction for the return date, published in the *Journal des scavans*, Clairaut appended the oddly prophetic remark, that a body travelling in regions so remote, as to be invisible for long periods, might be subject to totally unknown forces, perhaps the action of other comets, even maybe of some planet so distant from the central luminary as to be imperceptible.[1]

As to be expected, this aroused no interest at the time, being received for what it really was—pure conjecture. But six decades or so later, as the balance of informed opinion began to shift in favour of the trans-Uranian planet, after Bouvard's discomfiture in 1820–1 over the residual irregularities in the motion of Uranus, Clairaut's suggestion assumed a deeper and more relevant significance and was echoed with greater frequency first by one then another, until in November 1834 the English amateur astronomer, the Reverend Dr Thomas John Hussey, Vicar of Hayes in Kent, actually proposed in correspondence with George Airy to undertake a search for the hypothetical body—a project from which he was dissuaded by some adverse comment by Airy.

Yet as Hussey quietly discarded his deflated enthusiasm, surprising news drifted north—the offending planet had at last been sighted! At least according to Niccolo Cacciatore (1780–1841) Director of the Palermo Observatory, Sicily. 'One important thing I must communicate to you,' he reported to Capt W. H. Smyth, Foreign Secretary of the Royal Astronomical Society, late in 1835:

In the month of May (1835) I was observing the stars that have proper motion; a labour that has employed me several years. Near the 17th star, 12th hour, of Piazzi's Catalogue, I saw another, also of the 7.8th magnitude, and noted the approximate distance between them. The weather not having permitted me to observe on the two following nights, it was not till the third night that I saw it again, when it had advanced a good deal, having gone further to the eastward and towards the equator. But clouds obliged me to trust to the following night. Then, up to the end of May, the weather was horrible; it seemed in Palermo as if winter had returned: heavy rains and impetuous winds succeeded each other, so as to leave no opportunity of attempting anything. When at last the weather permitted observations at the end of a fortnight, the star was already in the evening twilight, and all my attempts to recover it were fruitless: stars of that magnitude being no longer visible. Meanwhile the estimated movement, in three days, was 10″ in A.R., and about a minute, or rather less, towards the north. So slow a motion would make me suspect the situation to be beyond *Uranus*. I was exceedingly grieved at not being able to follow up so important an examination.[2]

Smyth sent this report to Airy, President of the Society, but though it appeared in the *Monthly Notices* it drew little response from British astronomers. Meanwhile it had been translated by a Capt Basil Hall and despatched to the French Academy of Sciences. When published in the proceedings of that august body for February 15 1836, it evoked, by way of contrast to its London reception, some very interesting comments indeed, especially from a Louis François Wartmann (1793–1864), astronomer at the Geneva observatory.

Writing to François Arago at Paris, Wartmann recalled how some four and a half years before, he also had seen a moving star, though in a different part of the heavens to that described by Cacciatore.

Apparently, so the story goes, things had been difficult at Geneva during the summer of 1831. The Director, Professor Gautier, had retired to his château at Vinzel while the old observatory was being demolished to make way for a new structure. Equipment on order from Paris had not yet arrived, and the principal observatory instruments had been dismantled and placed in storage, with the exception of two small telescopes; a small Fraunhofer refractor and a 7° field Cauchoix finder which gave a magnification of 10.

Undaunted by his enforced idleness and the disordered state of things, Wartmann decided to prepare for his observations of Uranus; and early in September, with the Cauchoix finder, began a systematic series of observations on faint telescopic stars in Capricornus, from which he constructed a rough field chart of those situated along the route to be followed by the planet.

Conditions were excellent on the night of September 6 1831. The Moon had set, the sky perfectly cloudless, as Wartmann examined the relevant part of Capricornus, comparing it with his chart. After some time spent in this occupation, it seemed to him that one star, of pallid hue and about the 7th or 8th magnitude, had shifted slightly to the west, that is, in the retrograde direction. Significantly, it did not scintillate like its neighbours. Thinking perhaps error had been introduced into his original observations, Wartmann noted to re-observe the star at the next opportunity. Unfortunately the weather broke, and a fortnight or so elapsed before he again saw it. But on September 25 the cloud dispersed, and under clear skies with atmospheric vibration at a minimum the Cauchoix finder was trained on the southern part of Capricornus, where the suspect glistened not far from star 481, Hour XX of Piazzi's Catalogue, fainter than on the 6th, and of a golden shade. It had drifted still further towards the Sun. Assured on this point, Wartmann attempted to resolve it as a disk with a power of 60 on the Fraunhofer refractor, but without certain result. Miserable weather descended, but when the skies cleared on October 15, the Moon, then in its first quarter, and located in Capricornus, drowned out the fainter stars and Wartmann had some difficulty in finding the moving star. Eventually it was seen, much dimmer (about magnitude 9 or 10), and glowing with a yellow light. Again he confirmed its western drift. Wartmann effected his last positive identification on November 1 1831, when he found it still marching towards the Sun, and markedly orange in colour. He compared its brightness with that of 443, Hour XX of Piazzi's Catalogue, and thought it much fainter than when last seen.

Shortly after this observation, cloud, haze and fog brought the series to a close. Though he did obtain an occasional fleeting glimpse of the star, Wartmann failed to make any definite observations, and by the time the weather improved it had vanished into the twilight.

Wartmann had, of course, notified his superior of the observations, and the latter had offered to return, but the onset of inclement weather obliged him to defer this move. Additionally, Wartmann had also informed the Baron Franz Xavier von Zach (1754–1832) at Paris, on October 22 1831, and had sent him a copy of his chart. Apart from expressing a keen interest in the discovery, serious illness prevented the Baron from taking any effective action to assist in its confirmation. Since the latter died in the following year, Wartmann's letter and chart passed with the rest of the Baron's papers and library into the hands of the Baron Bernhard von Lindenau (1780–1854), no doubt unnoticed, and subsequently lost. A fact on which Wartmann sadly reflected in his notice to Arago.

As to the character of the moving star Wartmann made three proposals. First, he drew a comparison with the new star observed by Tycho Brahe in 1572, which is rather odd since this was immobile whereas Wartmann's object was not; this gives a rather curious sidelight on the man himself. Second, the object might have been a comet, but he disinclined to favour this hypothesis. Third and finally, the idea Wartmann did support:

> ... It would seem most probable that this minute point is a planet which travels around the Sun in an orbit of considerable radius. ... The new planet must needs lie at about double the distance of Uranus from the Sun, i.e., at a radius of 388, that of the Earth being 10; its period of revolution must therefore be about 243 years.'[3]

Both Cacciatore and Wartmann came in for sharp criticism, especially from Olbers and Valz, who thought the observations totally inadequate on which to raise a trans-Uranian hypothesis. Valz was suspicious that motion alone would be the determining factor; whilst Olbers suggested tests whereby the nature of the reported bodies could be discovered. Nothing came of these reflections. Yet in 1847 the matter still haunted the astronomical corridors. Hind wrote in the *Monthly Notices* of that year:

> I have been puzzling myself about Wartmann's supposed planet of 1831; and the only conclusion I can arrive at is, that no orbit will represent the four observations given in the *Comptes Rendus*. Professor Walker, of the Washington Observatory, has come to the same conclusion. There is something strange about the whole matter.[4]

And so we may agree today. These observations are strange and defy explanation. The four observations to which Hind refers are reproduced here:

Observed Positions of Wartmann's Planet 1831[5]

Date	d	h	m	RA °	RA ′	Dec °	Dec ′
September	6	22	30	315	27	17	28
September	25	19	00	315	09	17	42
October	15	20	00	314	52	17	51
November	1	20	30	314	36	17	59

If our disposition accords us the luxury of validating these observations, we are left with a definite enigma. But the truth is, they are too slight and much too whimsical to authenticate their mention on any other grounds except those of mere interest. To regard them in any other way would be to mislead. Admittedly we stand in correction on this point, but it must be remarked that these observations were not confirmed, and that the refinements expected of the professional observer, or indeed the advanced amateur, were not applied, at least they are not in evidence. Hence it is reasonable to infer that large observational errors are present, and if this is so, then the reality of the whole is countered and the explanation is rather more prosaic than either Cacciatore or Wartmann supposed.

We may conclude by mentioning that in the fourth volume of the *Marktree Catalogue* will be found a list of objects originally enumerated as stars, but found from later observation to be missing. The list is impressively long, and as its author remarked, possibly implied several unknown minor planets.[6] Missing stars find a place in Smyth's *Speculum Hartwellianum* for those interested to follow up the subject.[7]

Notes and References

Chapter I   The Search for a Satellite of the Moon

1. Moulton, F. R. *Astronomy* (1931), 199
2. Pickering, E. C. *Harvard Annals* (1890), 77
3. Barnard, E. E. *Astrophysical Journal* (1895), 347
4. Schafarik, C. V. *Astronomical Register* (1885), 23, no 273, 208
5. Pickering, E. C. *Harvard Annals* (1890), 77–8
6. Tombaugh, C. W. *The Search for Small Natural Earth Satellites.* *Final Technical Report* (1959) 97–8
7. Pickering, E. C. *Harvard Annals* (1890), 78–9
8. Tombaugh, C. W. *Search for Natural Satellites of the Earth by an Optical Technique.* *Quarterly Status Report* no 5 (1957), 1
9. Ibid, 2
10. Haas, W. *Popular Astronomy* (1943), 51, 397–400
11. Originally created by Barbara Middlehurst, Lunar and Planetary Laboratory, Tucson. Supported by NASA and the Smithsonian Institution Center for Short-Lived Phenomena. Stood down after Apollo 13
12. Tombaugh, C. W. *Final Report* (1959), 86
13. Ibid, 96
14. Ibid, 86
15. Pickering, E. C. *Harvard Annals* (1890), 79–80
16. Tombaugh, C. W. *Final Report* (1959), 86
17. Pickering, E. C. *Harvard Annals* (1890), 77
18. Ibid, 80
19. Ibid
20. Ibid, 79–80
21. Ibid, 81–2
22. Ibid, 82
23. Ibid, 82
24. Ibid, 82
25. Ibid, 82–3

26. Ibid, 82
27. *Annual Report of the Director of the Astronomical Observatory of Harvard College* (1888), 6
28. Willy Ley in his *Watchers of the Skies* (1964), page 262, quotes W. H. Pickering: 'On a Photographic Search for a Satellite of the Moon', as having been published in *Popular Astronomy*, 1903. The reference is incorrect. No such paper can be traced in that location
29. Barnard, E. E. *Astrophysical Journal* (1895), 2, 349
30. Ibid, 347
31. Ibid, 348
32. Ibid
33. Ibid
34. See Tombaugh, C. W. Reminiscences of the Discovery of Pluto. *Sky and Telescope* (1960), 19, no 5
35. See, T. J. J. *Astronomische Nachrichten* (1909), 181, 345–6
36. Tombaugh, C. W. *The Search for Small Natural Earth Satellites; Final Technical Report*, 30 June 1959, 2
37. Ashbrook, J. *Sky and Telescope* (1955), 14, no 6, 236
38. Pickering, W. H. *Popular Astronomy* (1923), 31, 23
39. Tombaugh, C. W. *Final Technical Report* (1959), 3
40. Ibid
41. Ibid
42. Ibid, 4
43. Ibid, 3
44. Ibid
45. Ibid, 5 The 'off-set' technique was not used in the case of the Moon satellite search, as it was thought better to reach a fainter magnitude instead of dividing the image in two
46. Tombaugh. *Final Technical Report* (1959), 95
47. Tombaugh. *Quarterly Status Report* no 5 (1957), 4. Also private communication Tombaugh to Baum, January 21, 1971

Chapter 2   *The Himalayas of Venus*

1. Herschel, J. F. W. *Outlines of Astronomy* (1850), 311–12
2. Schröter, J. H. *Phil Trans*, 85, 120 (1795)
3. Ibid
4. Ibid, 121
5. Ibid
6. Schröter, J. H. *Aphrod Frag* (1796), 13–15
7. Ibid, 29–32

8. Schröter, J. H. *Phil Trans*, 82, 315 (1792)
9. Ibid., 316–19
10. Ibid, 322
11. Ibid, 322–3
12. Ibid, 323–4
13. Ibid, 312
14. Herschel, W. *Phil Trans*, 83, 216 (1793)
15. Schröter, J. H. *Phil Trans*, 82, 333 (1792)
16. Ibid, 335
17. Ibid, 334
18. Ibid, 336–7
19. Herschel, W. *Phil Trans*, 83, 202 (1793)
20. Ibid
21. Ibid, 212
22. Ibid, 211
23. Ibid, 215
24. Ibid, 215–16
25. Ibid, 216
26. Schröter, J. H. *Phil Trans*, 85, 124 (1795)
27. Ibid, 117
28. Ibid, 119, 121
29. Ibid, 125
30. Ibid, 120
31. Ibid, 123–4
32. Ibid, 126–49
33. Ibid, 133–4
34. Ibid, 150
35. Ibid, 156
36. It is doubtful if his micrometers were of sufficient delicacy; indeed it was well known that his values of the diameters of certain minor planets were much too great
37. Breen, J. *The Planetary Worlds* (1854), 154–6
38. Porro, F. *JBAA*, 3, 184 (1893)
39. Alexander, W. *JBAA*, 3, 233 (1893)
40. Chambers, G. F. *Handbook of Astronomy*, I (1889), 100
41. Denning, W. F. *MNRAS*, 42, 111 (1882), Clerke, *History of Astronomy* (1885), 297–8, associated this feature with that delineated by De Vico in 1841. Certainly there are points of resemblance, but the matter is too uncertain to infer positive identification. At any rate the practical attitude adopted by Denning, who was keenly sceptical of all such appearances, does suggest that an objective configuration was seen

42. Denning, W. F. *Astr Reg*, 11, 131 (1874)
43. Sargent, F. *English Mechanic*, no 2511, 331 (1913)
44. McEwen, H. *JBAA* 23, 325–7 (1913)
45. Ibid, 326
46. Schröter, J. H. *Phil Trans*, 85, 145 (footnote) (1795)
47. Antoniadi, E. M. *MNRAS*, 58, 313–20 (1898)
48. Herschel, W. *Phil Trans*, 83, 218 (1793)
49. Gruithuisen, F. von P. *Bull des Sci math*, 1, 207, also in *Nova Acta Acad. Naturae Curiosorum*, 10, 239 (plate 19) (1820). Referred to by Clerke, *History of Astronomy* (1885), 301, and Trouvelot, *Obs sur les plan. Vénus et Mercure* (1892), 87
50. Vögel and Lohse. *Bothkamp Beobachtungen*. heft 2, 120 (1872–5)
51. Schiaparelli, G. V. *Ciel et Terre*, 11, 1890–1. Translation of his original paper from *Rendiconti del R. Istituto Lombardo*. 23, Serie ii (1890). See *Astr. Nach.* 138, 249–50 (1895)
52. Lowell, P. *The Evolution of Worlds* (1909), 77
53. Niesten, L., and Stuyvaert, E. *Obs sur L'aspect physique de Vénus de 1881 a 1895* (Bruxelles, 1903)
54. Lohse and Wigglesworth. *MNRAS*, 47, 495 (1886)
55. Denning, W. F. *Telescopic Work for Starlight Evenings* (1891), 148
56. Percival, P. *The Evolution of Worlds* (1909), 77. These observations were made with the 24-inch objective at Flagstaff
57. Cragg, T. L. Private communication Nov 1953. See *The Strolling Astronomer*, May 1953, 69
58. Holden, A. P. *Astr Reg*, 8, 118 (1871)
59. Trouvelot, E. L. *BSAF*, 6, 61–147 (1892)
60. Trouvelot, E. L. *The Observatory*, 3, 416–17 (1880), and *CR*, 98, 719 (1884)
61. Trouvelot, E. L. *BSAF*, 6, 81 (1892)
62. Zenger, C. V. *MNRAS*, 37, 461 (1877)
63. Ibid
64. Langdon, R. *MNRAS*, 33, 500–01 (1873)
65. Pratt, H. *Astr Reg*, 10, 43 (1873)
66. Denning, W. F. *Astr Reg*, 11, 131 (1874)
67. Denning, W. F. *MNRAS*, 42, 111 (1882)
68. Grye, B. de la. *CR*, 98, 1406 (1884). '. . . probably due to "photographic irradiation" from a local excess of brilliancy, the result—according to the French investigators' conjecture—of accumulations of ice and snow, or the continuous formation of vast cloud-masses.' Clerke, *History of Astronomy* (1885), 301
69. Denning, W. F. *Telescopic Work for Starlight Evenings* (1891), 148
70. Ibid

71. Lohse, I. G. *MNRAS*, 47, 495 (1886)
72. Bartlett, J. C. *The Strolling Astronomer* 7, 70 (1953)
73. Mascari, A. *Astr Nach*, 139, 257–64 (1896)
74. Zona, T. *Astr Nach*, 131, 121–4 (1892)
75. Niesten, L. and Stuyvaert, E. *Obs sur L'aspect physique de Vénus de 1881 à 1895* (Bruxelles, 1903)
76. Trouvelot, E. L. *BSAF*, 6, 96–7 (1892)
77. Phase anomaly; the acceleration of phase in eastern elongation, dichotomy being from 4 to 8 days sooner than the theoretical time, and the corresponding retardation at western elongation, rather more pronounced, though once considered due to mountains (Schröter) has not been included. It is conceivably due to other factors somewhat divorced from the mountain hypothesis
78. Sagan, C. *Nature*, 216, 1198–9 (1967)
79. Dollfus, A. *Planets and Satellites* (1961), 553
80. Goldstein, R. M., and Rumsey, H. *A Radar Snapshot of Venus.* Science, 169, 974–7 (1970). See, Jurgens, R. F. *Radio Science*, 5, 435–42 (1970). Rogers, A. E. E., and Ingalls, R. P. *Radio Science*, 5, 425–33 (1970). Smith, W. B., Ingalls, R. P., Shapiro, I. I., and Ash, M. E. *Radio Science*, 5, 411–23 (1970). Evans, J. V. 'Radar Studies of Planetary Surfaces'. *Annual Review of Astronomy and Astrophysics* 7 (1969). Carpenter, R. L. Study of Venus by cw Radar—1964 Results. *Astr Jnl*, 71, 142–52 (1966)
81. Wheelock, H. (compiler). *Mariner: Mission to Venus* (New York, 1963), 105–06

Chapter 3 *'An Unexplained Observation'*

1. Newcomb, Simon. Reminiscences of an Astronomer (1903)
2. Prize offered by an American businessman H. H. Warner for every new comet found by observers in the United States or Canada. It amounted to 200 dollars for each discovery. Discontinued, then revived in 1890 by another wealthy American, J. A. Donohoe. See Chambers: *The Story of the Comets* (1909), page 55
3. Holden, Edward S. *Handbook of the Lick Observatory* (San Francisco, 1888)
4. Barnard, E. E. *Astr Jnl.* 12, 81–5 (1892)
5. Barnard, E. E. *Astr Nach*, 173, 316–17 (1907)
6. Barnard, E. E. *Astr Nach*, 173, 315–16 (1907). First part given by Barnard direct from his notebook. Second part incorporated from his report of the observation, *Astr Nach*, 172, 26 (1906)

7. Barnard, E. E. *Astr Nach*, 173, 315–16 (1907). He notes from his workbook, 'No small star near it [Venus] except the one preceding, . . . Examined Venus from 16h 30m to 16h 50m, . . . Star 1′ south, star 6/10 of 260 field (field = 6′) about 14s.'
8. Barnard, E. E. *Astr Nach*, 173, 315 (1907)
9. Barnard, E. E. *Astr Nach*, 172, 25 (1906)
10. Ibid, 26
11. Ibid
12. Barnard, E. E. *Astr Nach*, 172, 25–26 (1906)
13. Pirovano, R. *Astr Nach*, 172, 207–08 (1906)
14. Barnard, E. E. *Astr Nach*, 173, 315, 317 (1907)
15. Ibid, 318
16. Ashbrook, J. *Sky and Telescope*, 15, 356 (1956)
17. Ibid

Chapter 4   A Strange Celestial Visitor

1. *HCO Bulletin*, no 757
2. Campbell, W. W. *PASP*, October 1921, 258
3. Ibid
4. Ibid, 258–9
5. Ibid, 259
6. Ibid
7. *Astr Nach*, 214, 69 (1921)
8. Fellows, S. *English Mechanic*, no 2943, 49 (1921)
9. *Nature*, 108, 69 (1921)
10. Wolf, M. *Astr Nach*, 214, 103 (1921)
11. Gaythorpe, S. B. *JBAA*, 32, 67 (1921)
12. Ibid
13. Ibid
14. Crommelin, A. C. D. *JBAA*, 32, 46 (1921)
15. Gaythorpe, S. B. *JBAA*, 32, 67 (1921)
16. *HCO Bulletin*, no 759
17. Campbell, W. W. *PASP*, October 1921, 261
18. Crommelin, A. C. D. *JBAA*, 31, 367 (1921)
19. Campbell, W. W. *PASP*, October 1921, 262
20. *Nature*, 108, 69 (1921)
21. Campbell, W. W. *PASP*, October 1921, 260
22. Ibid, 262
23. Ibid
24. *English Mechanic*, no 2943, 47 (1921)
25. *Nature*, 108, 69 (1921)

26. Wolf, M. *Astr Nach*, 214, 69–70 (1921)
27. *English Mechanic*, no 2943, 49 (1921)
28. Campbell, W. W. *PASP*, October 1921, 260
29. Ashbrook, J. *Sky and Telescope*, 41, 353 (1971)

Chapter 5   William Herschel and the Ring of Uranus

1. Occurs in a letter by Herschel to Lichtenberg, and published in German in Lichtenberg and Forster; *Göttunger Magazin der Wissenschaften und Literatur*, 3, 4. Letter from Datchet, near Windsor, dated November 15 1783. English version appears in Holden; *Sir William Herschel, His Life and Works* (New York, 1881), 4
2. Herschel, W. *Phil Trans*, 77, 125–6 (1787)
3. Ibid, 126
4. Ibid
5. Ibid
6. Ibid, 128
7. Herschel, W. *Phil Trans*, 88, 67 (1798)
8. Ibid
9. Ibid
10. Ibid
11. Ibid
12. Ibid
13. Ibid, 67–8
14. Ibid, 68
15. Ibid
16. Ibid
17. Ibid, 69
18. South, J. *MNRAS*, 1, 181 (1828–33)
19. Herschel, J. F. W. *MNRAS*, 3, 36 (1834)
20. In Denning: *Telescopic Work* (1891), 28
21. Herschel, W. *Phil Trans* 88, 70 (1798)
22. Ibid. Humboldt: *Cosmos 4* (1852, Bohn ed.), 525. Sabine ed. *3* (1852), 388, remarked; 'The original supposition that Uranus had two rings was found to be an optical illusion by the discoverer himself, in all cases so cautious and persevering in confirming his discoveries.'

Chapter 6   Neptune 1846–7

1. Grosser, M. *The Discovery of Neptune* (Harvard, 1962) and Jones,

Sir Harold Spencer; *John Couch Adams and the discovery of Neptune* (Cambridge, 1947), are amongst the best contemporary versions of this celebrated event

2. Huggins, W. *MNRAS*, 41, 188–91 (1881)
3. James Nasmyth co-authored with James Carpenter the classic *The Moon: Considered as a Planet, a World, and a Satellite* (1874). Advocated volcanic origin of lunar surface details
4. Herschel, J. F. W. *MNRAS*, 9, 88 *et seq* (1849)
5. Ibid
6. Hind, J. R. *The Times*, October 1 1846, page 8
7. Lassell, W. *The Times*, October 14 1846, page 7. Quoted by Smyth: *Cycle of Celestial Objects* (1860), 416
8. Lassell, W. *MNRAS*, 7, 167–8 (1846). Noted in Annual Catalogue of Papers 1846–7, *MNRAS*, 7, 235 (1847)
9. Leverrier placed Neptune in Capricorn, 5° east of delta Capricorni, whereas according to modern configurations it lay in Aquarius. In 1846 constellation boundaries were not internationally agreed, and varied from chart to chart
10. Lassell to Challis, January 19 1847. Letter. (43)
11. Lassell, W. *MNRAS*, 7, 168 (1846)
12. Lassell to Challis, January 19 1847. Letter. (43)
13. Dawes to Challis, April 7 1847. Letter. (44)
14. Lassell, W. *Astr Nach*, 26, 165–6 (1848)
15. Smyth, W. H. *Cycle of Celestial Objects* (1860) 416
16. Ibid, footnote page 417
17. Breen, James. *The Planetary Worlds* (1854) 249–50
18. Lassell, W. *MNRAS*, 7, 157 (1846). Annual Catalogue of Papers February 1846 to February 1847, *MNRAS*, 7, 232–6 (1847)
19. Ibid
20. Lassell, W. *MNRAS* 7, 167–8 (1846)
21. Ibid, 168
22. Hind, J. R. *MNRAS*, no 589, 207–08 (1847)
23. Lassell to Challis, January 19 1847. (43)
24. Mitchel, O. M. *The Sidereal Messenger*, 1, 60 (1846)
25. Walker, R. L. Private communication March 1 1966. Maury's biographies apparently contain no reference to this observation
26. Hind, J. R. *The Solar System* (1851), 136–7
27. Bond, W. C. *MNRAS*, 8, 9 (1848), footnote
28. Ashbrook, J. Private communication August 7 1970
29. Rawlins, D. Private communication October 22 1970
30. Challis, James. *Astr Nach*, no 596. Reprinted in *The Sidereal Messenger*, 1, 113–14 (1847)

31. Challis to Lassell, January 16 1847. (Letter) Mentioned by Challis, reference 30 above. Lassell's letter of January 19 is the reply
32. Lassell to Challis, January 19 1847
33. Challis, J. *Astr Nach*, 25, 231-2 (1847)
34. Dawes to Challis, April 7 1847
35. *MNRAS*, 7, 217 (1847)
36. Shapley, H. *Sky and Telescope*, 11, 160 (1952)
37. Dawes to Challis, April 7 1847
38. Ibid
39. Challis, James. Repetitive comment. Whether he submitted basically the same report, or whether his sensitive position at that time made him self-conscious of having missed another discovery is not clear
40. Lassell to Challis, January 19 1847. See also *MNRAS*, 7, 167-8 and *Astr Nach*, 26, 167-8 (1848) with regard to Lassell's confidence in the use of his Merz prism as a secondary instead of a flat
41. Grant, R. *History of Physical Astronomy* etc (1852), 190. Original, Herschel, Sir John F. W. Le Verrier's planet (letter), *Athenaeum*, October 3 1846, 1019
42. Humboldt, Alexander von. *Cosmos* (4 vols) Otté. (1849–1852), vol 4, 476, footnote

Chapter 7   Strange Interlude

1. *Publ Lick Obs*, 19 (1935), page 14
2. *Washington Obs*, 1850, Introduction, page xxx
3. Ferguson. *Astr Jnl*, 1, 165 (1851)
4. Hind. *Astr Jnl*, 2, 78 (1852)
5. Ibid
6. Ferguson. *Ast Jnl*, 1, 166 (1851)
7. Hind. *Astr Jnl*, 2, 78 (1852)
8. Ferguson. *Astr Jnl*, 164 (1851)
9. Ibid, 165
10. Maury. *Astr Jnl*, 2, 53 (1851)
11. Ibid
12. Ibid
13. Maury. *Astr Jnl*, 2, 91 (1852)
14. Maury. *Astr Jnl*, 2, 53 (1851)
15. Hind. *Astr Jnl*, 2, 78 (1852)
16. Maury. *Astr Jnl*, 2, 91 (1852)
17. Hind. *Astr Jnl*, 2, 78 (1852)

18. Ibid
19. Ibid
20. Maury. *Astr Jnl*, 2, 91 (1852)
21. Ibid
22. Ibid. Col A, in the first table indicates state of the atmosphere, 10 being the most favourable
23. Lockyer. *Nature*, 18, 461 (1878)
24. Anon. *Nature*, 18, 696 (1878)
25. Peters. *AN*, 94, cols 115–16 (1879)
26. Ibid
27. Ibid
28. Rodgers. *AN*, 94, cols 113–14 (1879)
29. Zinner. *Astron Abhandl*, 8, A35 and A75 (1931)
30. Keats. *On first looking into Chapman's 'Homer'*

Chapter 8   The Lost Planets of 1831 and 1835

1. Clairaut, A. C. *Journal des scavans* (1759), 86
2. Smyth, W. H. *MNRAS*, 3, 139 (1835)
3. Wartmann, L. F. *Comptes Rendus*, 2, 311 (1836)
4. Hind, J. R. *MNRAS*, 7, 274 (1847)
5. Wartmann. 307
6. Cooper, E. J. *Catalogue of Stars near the Ecliptic*. 4 vols (Dublin, 1851–6). Contains 60,066 stars observed at the Marktree Observatory, Ireland between 1848–56
7. Smyth, W. H. *The Cycle of Celestial Objects* (1860), 236–7

Cacciatore's notice appeared in translation in the *Comptes Rendus*, 2, 154–5 (1836). Olbers' note in the same journal, 3, 141 (1836)

Bibliography

Chapter 1   The Search for a Satellite of the Moon

Barnard, E. E. 'On a Photographic Search for a Satellite to the Moon', *The Astrophysical Journal*, 2, 347–9 (1895)

Baum, R. M. 'A Meteoric Satellite to the Moon', *The Strolling Astronomer*, 10, 7–11 (1956)

Baum, R. M. 'Is there a Satellite to the Moon?', *Vega*, 31, March 16 (1956)

Haas, W. H. 'Concerning Possible Lunar Meteoric Phenomena', *Popular Astronomy*, 51, 397–400 (1943)

Moulton, F. R. *Astronomy* (New York, 1931)

Pickering, E. C. 'Photographic Search for a Lunar Satellite', *Miscellaneous Researches made during the years 1886–1890. Annals of the Astronomical Observatory of Harvard College*, 18, no 4, 77–83 (Cambridge, Mass 1890)

Pickering, E. C. *Forty-third Annual Report of the Director of the Astronomical Observatory of Harvard College*. Presented to the Visiting Committee December 15 1888, page 6 (Cambridge, Mass 1888)

See, T. J. J. 'Dynamical Theory of the Capture of Satellites and of the division of nebulae under the secular action of a resisting medium. VIII. Theoretical and Observed Distances of Satellites in the Solar System', *Astronomische Nachrichten*, 181, nos 4341–2 (1909)

Tombaugh, C. W. 'Search for Small Satellites of the Moon during the Total Lunar Eclipse of November 18 1956', *The Strolling Astronomer*, 11, 61–4 (1957)

Tombagh, C. W., Smith, B. R., and Capen, C. F. 'Search for Small Satellites of the Moon during the Total Lunar Eclipse of November 18 1956' (Abstract), *Publications of the Astronomical Society of the Pacific*, 69, 400–01 (1957)

Tombaugh, C. W. 'Search for Natural Satellites of the Earth by an Optical Technique', *Quarterly Status Report No. 5*, Period October 15 1956 to January 15 1957. (Office of Ordnance Research Project no 1602, Contract no DA-04-495-ORD-727.) New Mexico College of Agriculture and Mechanic Arts, State College, New Mexico (1957)

Tombaugh, C. W., Robinson, J. C., Smith, B. A., and Murrell, A. S. *The Search for Small Natural Earth Satellites. Final Technical Report, 30 June 1959. Pt. 5 Search for Small Satellites of the Moon during the Total Lunar Eclipse of November 18, 1956.* 86–98, New Mexico State University, Physical Science Laboratory (1959)

Miscellaneous material

Chant, C. A. 'Meteors on the Moon', *Journal of the Royal Astronomical Society of Canada*, 37, 216 (1943)

Ley, W. *Watchers of the Skies. An Informal History of Astronomy from Babylon to the Space Age* (1964)

Ley, W. *Rockets, Missiles and Space Travel* (New York, 1951)

Schafarik, C. V. 'Telescopic Meteors', *The Astronomical Register*, 23, no 273, September 1885

Watson, F. G. *Between the Planets* (1948). Also the revised paperback edition (Doubleday 1962)

Whipple, F. L. 'The meteoritic environment of the Moon', *Proceedings of the Royal Society, A*, vol 296, 304–15 (1967)

Chapter 2 *The Himalayas of Venus*

Alexander, W. 'Venus' (letter), *Journal of the British Astronomical Association*, 3, 233 (1893)

Antoniadi, E. M. 'Notes on the Rotation Period of Venus', *Monthly Notices of the Royal Astronomical Society*, 58, 313–320 (1898)

Bartlett, Jr, J. C. 'The Apparent South Pole of Venus with some notes on the Inclination', *The Strolling Astronomer*, 7, 65–72 (1953)

Breen, J. *The Planetary Worlds; the topography and telescopic appearances of the Sun, Planets, Moon, and Comets* (London, 1854), 154–6

Denning, W. F. 'Venus, Jupiter's Satellites' (letter), *The Astronomical Register*, 11, 130–2 (1874)

Denning, W. F. 'Observations of Venus in the Spring of 1881', *Monthly Notices of the Royal Astronomical Society*, 42, 109–12, (1882)

Denning, W. F. *Telescopic Work for Starlight Evenings* (London, 1891)

Ertborn, O. van. 'Observations de la planete Vénus en 1876', *Bullétin de l'Académie Royale de Sciences, des Lettres, des Beaux-Arts de Belgique. 2ème Série*, 43, 20–4 (1877)

Herschel, W. 'Observations on the Planet Venus', *Philosophical Transactions*, 83, 201–19 (1793)

Holden, A. P. 'The Planet Venus' (extract), *The Astronomical Register*, 8, 118 (1871)

Langdon, R. 'Star-spots on Venus', *Monthly Notices of the Royal Astronomical Society*, 33, 500–01 (1873)

Lowell, P. *The Evolution of Worlds* (1909)

Lowell, P. 'Detection of Venus' Rotation Period and of the Fundamental Physical Features of the Planet's Surface', *Popular Astronomy*, 4, 281–5 (1896)

Mascari, A. 'Osservazioni del planeta Venere', *Astronomische Nachrichten*, 139, 257–64 (1896)

McEwen, H. 'Large Irregularity on the Terminator of Venus, February 1913', *Journal of the British Astronomical Association*, 23, 325–7 (1913)

Niesten, L., and Stuyvaert, E. 'Observations sur L'aspect physique de Vénus de 1881 à 1885', *Annales astronomique de l'Observatoire royal de Belgique*, 8, nouvelle série (1903)

Pratt, H. 'Markings on Venus' (letter), *The Astronomical Register*, 10, 43 (1873)

Porro, F. 'Observations of Major Planets at Turin Observatory', *Journal of the British Astronomical Association*, 3, 184 (1893)

Quénisset, F. 'Photographies de la planète Vénus', *Comptes Rendus des Séances de l'Académie des Sciences*, 172, 1645–7 (1921)

Ross, F. E. 'Photographs of Venus', *Astrophysical Journal*, 68, 57–92 (1928)

Sagan, C. 'Life on the Surface of Venus', *Nature*, 216, 1198–9 (1967)

Sargent, F. 'Venus' (letter), *English Mechanic*, no 2511, 331 (1913)

Schiaparelli, G. V. 'Considerations sur le mode de rotation de la planète Vénus', *Ciel et Terre*, 9, 49–62, 125–43, 183–90, 214–222 (1890–1)

Schiaparelli, G. V. 'Zwei Schreiben von Herrn. Prof. G. Schiaparelli in Mailand an den Herausgeber betr. die auf der Oberfläche der Vénus beobachteten Flecken', *Astronomische Nachrichten*, 138, 249–52 (1895), and 138, 251–2, a further communication

Schiaparelli, G. V. 'Sur une tache récemment observée à la surface de Vénus et sur la duree de rotation de cette planète', (Extrait d'une lettre de M. Schiaparelli, associé de l'Academie, à M. Terby.) *Bulletin de l'Academie Royale de Belgique*, 3rd série, 30, 204–05 (1895)

Schröter, J. H. *Aphroditographische Fragmente zur genauern Kenntnifs des Planeten Venus etc* (Helmstedt, 1796)

Schröter, J. H. 'Observations on the Atmospheres of Venus and the Moon, their respective Densities, perpendicular Heights, and the Twilight occasioned by them', *Philosophical Transactions*, 82, 309–37 (1792)

Schröter, J. H. 'New Observations in further Proof of the mountainous Inequalities, Rotation, Atmosphere, and Twilight, of the Planet Venus.' *Philosophical Transactions*, 85, 117–176 (1795)

Trouvelot, E. L. 'White Spots on Venus' (letter), *The Observatory*, 3, 416–17 (1880)

Trouvelot, E. L. 'Les taches polaires de Vénus', *Comptes Rendus des Séances de l'Académie des Sciences*, 98, 719–20, and 1481–2, (1884)

Trouvelot, E. L. 'Observations sur les planètes Vénus et Mercure', *Bulletin Société Astronomique de France*, 6, 61–147 (1892)

Vögel, H. C., and Lohse, I. G. *Beobachtungen angestellt auf der Sternwarte des Kammerherrn von Bülow zu Bothkamp*, 3 parts (Leipzig, 1872–5)

Wheelock, H. J. (compiler). *Mariner Mission to Venus* (New York, 1963)

Zenger, C. V. 'Absorption of the Light of *Venus* by Dark Violet Glass Plates', *Monthly Notices of the Royal Astronomical Society*, 37, 460–2 (1877)

Zona, T. 'Osservazioni di Venere', *Astronomische Nachrichten*, 131, 121–4 (1892)

Interesting and informative accounts on this subject may be found in the undermentioned works:

Chambers, G. F. *A Handbook of Descriptive and Practical Astronomy*, vol 1 (1889)

Clerke, A. M. *A Popular History of Astronomy during the Nineteenth Century* (1885)

Moore, P. *The Planet Venus* (1956)

Webb, The Rev T. W. *Celestial Objects for Common Telescopes* (1881). Dover paperback edition of the 6th edition published in 1917 (1962)

Chapter 3 'An Unexplained Observation'

Ashbrook, Joseph. 'Barnard's "Unexplained Observation" ', *Sky and Telescope*, 15, 356 (1956)

Barnard, E. E. 'An unexplained observation', *Astronomische Nachrichten*, 172, 25–6 (1906)

Barnard, E. E. 'Reply to Mr. Rudolph Pirovano's remarks in A.N. 4117, concerning "An unexplained observation" ', *Astronomische Nachrichten*, 173, 315–18 (1907)

Pirovano, Rudolf. 'Notiz. betr. E. E. Barnard, An unexplained observation, A.N. 4106', *Astronomische Nachrichten*, 172, 207–08 (1906)

Miscellaneous material

Barnard, E. E. 'Discovery and Observations of a Fifth Satellite to *Jupiter*', *The Astronomical Journal*, 12, 275, 81–5 (1892)

Chambers, G. F. *The Story of the Comets, simply told for general readers* (1909)

Holden, E. S. *Handbook of the Lick Observatory* (San Francisco, 1888)

Newcomb, S. *Reminiscences of an Astronomer* (1903)

Chapter 4 A Strange Celestial Visitor

Ashbrook, Joseph. 'W. W. Campbell and a Puzzling Object', *Sky and Telescope*, 41, 6, 352–3 (Cambridge, Mass 1971)

Bailey, S. I. 'Bright Object near the Sun', *Harvard College Observatory Bulletin* 757, August 9 1921 (Cambridge, Mass)

Bailey, S. I. 'Bright Object near the Sun', *Harvard College Observatory Bulletin* 759, October 14 1921 (Cambridge, Mass)

Campbell, W. W. 'Observations of an unidentified object seen near the Sun on Sunday, August 7 1921', *Publications of the Astronomical Society of the Pacific*, 32, 258–62, October 1921, (San Francisco)

Cook, Grace. 'A Luminous Night Sky: The Shadow of the Earth and a Sunset' (letter), *English Mechanic and World of Science*, vol 113–14, no 2943, 49, August 19 1921

Crommelin, A. C. D. 'Comet Catalogue. Containing the Orbits of Comets observed from the beginning of 1894 to April 1925, together with improved Orbits of many earlier Comets: being a sequel to Galle's Cometenbahnen', *Memoirs of the British Astronomical Association*, 26, 2, 21 (1925)

Crommelin, A. C. D. 'Report of the Comet Section 1920–1921' *Journal of the British Astronomical Association*, 31, 367–8 (1921)

Crommelin, A. C. D. 'Report of the November Meeting, 1921', *Journal of the British Astronomical Association*, 32, 46 (1921)

Fellows, S. 'New Star, or Comet, or What?', *English Mechanic and World of Science*, vol 113–14, no 2943, 49, August 19 1921

Gaythorpe, S. B. 'Note on a Luminous Body, 22° West of the Sun, seen before Sunrise on 1921 August 6', *Journal of the British Astronomical Association*, 32, 67, November 1921

Wolf, Max. 'Leuchtende Bänder am Nachthimmel', *Astronomische Nachrichten*, 214, no 5116, 69–70 (1921)

Miscellaneous material

'Neuer Komet in der Nähe der Sonne oder Nova' (telegram). *Astronomische Nachrichten* 214, no 5116, 69–70 (1921)

'Scientific News', *English Mechanic and World of Science*, vol 113–114, no 2943, 47, August 19 1921

'The Bright Object Near the Sun', *Nature*, 108, no 2706, 69, September 8 1921

Chapter 5 William Herschel and the Ring of Uranus

Herschel, W. 'An Account of the Discovery of Two Satellites revolving round the Georgian Planet', *Philosophical Transactions*, 77, 125–9 (1787)

Herschel, W. 'Observations and Reports tending to the Discovery of one or more Rings of the Georgian Planet, and the flattening of its Polar Regions', *Philosophical Transactions*, 88, 67–71 (1798). Forms part of a long paper 'On the Discovery of four additional Satellites of the Georgium Sidus. The retrograde Motion of its old Satellites announced; and the Cause of their Disappearance, at certain distances from the Planet explained'

Miscellaneous material

Alexander, A. F. O'D. *The Planet Uranus, A History of Observation, Theory and Discovery* (1965)

Ball, Sir R. S. *Great Astronomers* (1895)

Berry, A. *A Short History of Astronomy from the Earliest Times through the Nineteenth Century* (1898)

Buttmann, G. *The Shadow of the Telescope* (New York, 1970)

Hoskins, M. A. *William Herschel and the Construction of the Heavens* (1963)

Lubbock, C. (editor). *The Herschel Chronicle* (1933)

Chapter 6   Neptune 1846–7

MSS sources

The Rev W. R. Dawes. Letter to Professor James Challis, Plumian Professor of Astronomy, Cambridge, dated April 7 1847. Library of the Cambridge Observatories, England. (Accession no 44)

William Lassell. Letter to Professor James Challis, dated 1847 January 19. Library of the Cambridge Observatories, England (Accession no 43)

Articles

Bond, W. C. 'Notice of his observations on the satellite of Neptune', *Monthly Notices of the Royal Astronomical Society*, 8, 9 (1847)

Challis, J. 'Second Report of proceedings in the Cambridge Observatory relating to the new planet Neptune', *Sidereal Messenger*, 1, no 15, 113–14 (1847). Reprinted from the *Astronomische Nachrichten*, no 596

Challis, J. Abstract in 'Observations of Neptune; meridian observations from Cambridge and Hamburg', *Monthly Notices of the Royal Astronomical Society*, 7, 243–4 (1847)

Challis, J. Ring mentioned and illustrated in various meridian observations at Cambridge, *Astronomische Nachrichten*, 25, 231–2, (1847)

Hind, J. R. Notice of Lassell's observations and the South Villa study, *Astronomische Nachrichten*, 25, 207-08 (1847)

Huggins, Sir W. 'Obituary Notice of William Lassell', *Monthly Notices of the Royal Astronomical Society*, 41, 188-191 (1881)

Lassell, W. Abstract of 'Account of a Supposed Ring and Satellite of the New Planet in Observations of Le Verrier's Planet', *Monthly Notices of the Royal Astronomical Society*, 7, 157 (1846). See Annual Catalogue of Papers Read to the Royal Astronomical Society, *Monthly Notices of the Royal Astronomical Society*, 7, 232-6 (1847)

Lassell, W. 'Abstract of Account of Physical Observations of Le Verrier's Planet in Observations of Le Verrier's Planet', *Monthly Notices of the Royal Astronomical Society*, 7, 167-8 (1846). See Annual Catalogue of Papers Read to the Royal Astronomical Society, *Monthly notices of the Royal Astronomical Society*, 7, 235 (1847)

Lassell, W. 'On the discovery of a satellite of Neptune', *Monthly Notices of the Royal Astronomical Society*, 7, 297-8, 1847. See Annual Catalogue of Papers. *Monthly Notices of the Royal Astronomical Society*, 8, 98 (1848)

Lassell, W. 'Observations of Neptune and his Satellite', *Monthly Notices of the Royal Astronomical Society*, 7, 307-08 (1847). See Annual Catalogue of Papers. *Monthly Notices of the Royal Astronomical Society*, 8, 98 (1848)

Lassell, W. 'Schreiben des Herrn *Lassell* an den Herausgeber', *Astronomische Nachrichten*, 26, 165-8 (1848)

Lassell, W. 'Letter on "Le Verrier's Planet"', *The Times*, October 14 1846, page 7

Mitchel, O. M. 'The New Planet, Its Name and Discovery', *Sidereal Messenger*, 1, 60 (1846)

Shapley, H. 'Speculation on the possibility that other planets may have rings', *Sky and Telescope*, 11, 160 (1952)

Turner, H. H. 'Obituary Notice of Johan Gottfried Galle', *Monthly Notices of the Royal Astronomical Society*, 71, 275–81 (1911)

Books

Breen, J. *The Planetary Worlds: the topography and telescopic appearances of the Sun, Planets, Moon, and Comets* (London, 1854)

Chambers, G. F. *A Handbook of Descriptive and Practical Astronomy;* 3 vols (1889–90)

Denning, W. F. *Telescopic Work for Starlight Evenings* (1891)

Grant, R. *History of Physical Astronomy from the Earliest Ages to the Middle of the Nineteenth Century* (1852)

Grosser, M. *The Discovery of Neptune* (Harvard, 1962)

Herschel, Sir J. F. W. *Outlines of Astronomy*, 3rd ed (1850)

Hind, J. R. *The Solar System. A Descriptive Treatise Upon the Sun, Moon, and Planets, including an Account of all the Recent Discoveries* (1851)

Hind, J. R. *Astronomical Observations taken at the Observatory South Villa, Inner Circle, Regent's Park, London during the years 1839–1851* (1852)

Humboldt, A. Von. *Cosmos.* Translated by E. C. Otté, 5 vols (1849–58)

Jones, Sir H. Spencer. *John Couch Adams and the Discovery of Neptune* (1947)

Smyth, Vice-Admiral W. H. *The Cycle of Celestial Objects continued at the Hartwell Observatory to 1859. With a notice of recent discoveries, including details from the Aedes Harwellianae* (1860)

Chapter 7   *Strange Interlude*

Anon. 'A Missing Star', *Nature*, 18, 696, October 31 1878

Ashbrook, J. 'The Case of the Lost Trans-Neptunian Planet', *Sky and Telescope*, 15, 4, 169 (Cambridge, Mass, 1956)

Ferguson, J. 'Observations of Hygea. Made with the Filar-Micrometer of the Washington Equatorial', *The Astronomical Journal*, 1, 164–6 (1851)

Hind, J. R. 'Letter from Mr. Hind to the Editor', *The Astronomical Journal*, 2, 78 (1852)

Lockyer, J. N. 'The Eclipse', *Nature*, 18, 457–62, August 29 1878

Maury, M. F. 'Letter of Lieutenant Maury to Hon. William A· Graham, Secretary of the Navy', *The Astronomical Journal*, 2, 53 (1851)

Maury, M. F. 'Letter from Lieutenant Maury, Superintendent of the Washington Observatory', *The Astronomical Journal*, 2, 91 (1852)

Peters, C. H. F. 'Investigation of the evidence of a supposed trans-Neptunian planet in the Washington observations of 1850', to Rear-Admiral John Rodgers, USN, Superintendent

Naval Observatory. *Astronomische Nachrichten*, 94, nr 2240, cols 113–16 (1879)

Rodgers, J. 'Letter from Admiral John Rodgers, Superintendent of Naval Observatory at Washington', *Astronomische Nachrichten*, 94, nr 2240, cols 113–14 (1879)

Zinner, E. *Astronomische Abhandlungen. Ergänzungshefte zu den Astronomischen Nachrichten* 8 (1931)

Miscellaneous material

Leuschner, A. O. 'Research Surveys of the Orbits and Perturbations of Minor Planets 1 to 1091 from 1801.0 to 1929.5', *Contributions of the Berkeley Astronomical Department (Student's Observatory), University of California. Publications of the Lick Observatory*, vol 19 (1935)

Ley, W. *Watchers of the Skies. An Informal History of Astronomy from Babylon to the Space Age* (1964)

Chapter 8   The Lost Planets of *1831* and *1835*

Cacciatore, N. 'Sur une nouvelle petite planète dont l'existence a été soupçonnée par M. *CACCIATORE*, directeur de l'Observatoire de Palerme', *Comptes Rendus des Séances de l'Academie des Sciences*, 2, 154–5 (1836)

Hind, J. R. 'Wartmann's Supposed Planet'. *Monthly Notices of the Royal Astronomical Society*, 7, 274 (1847)

Smyth, W. H. 'Extract of a letter from Capt. Smyth to the President, containing the translation of a notice from M. Cacciatore', *Monthly Notices of the Royal Astronomical Society*, 3, 139 (1833)

Wartmann, L. 'Lettre de M. Wartmann, de Geneve, à M. Arago, sur un astre ayant l'aspect d'une étoile et qui cependant était doué d'un mouvement propre', *Comptes Rendus des Séances de l'Academie des Sciences*, 2, 307–11 (1836)

Miscellaneous material

Clairaut, A. C. 'Memoire sur la comete de 1682 addressee à MM. les auteurs de Journal des scavans par M. Clairaut', *Journal des scavans*, 80–96, Paris (1759)

Cooper, E. J. *Catalogue of Stars near the Ecliptic* (Dublin, 1851–6), 4 vols

Smyth, W. H. *The Cycle of Celestial Objects continued at the Hartwell Observatory to 1859* (1860). Also known by its spine title as the *Speculum Harwellianum*

Index